Angelo Heilprin

Explorations on the west Coast of Florida and in the Okeechobee

Wilderness

Angelo Heilprin

Explorations on the west Coast of Florida and in the Okeechobee Wilderness

ISBN/EAN: 9783337140021

Printed in Europe, USA, Canada, Australia, Japan

Cover: Foto ©ninafisch / pixelio.de

More available books at **www.hansebooks.com**

EXPLORATIONS

ON THE

WEST COAST OF FLORIDA

AND IN THE OKEECHOBEE WILDERNESS.

With Special Reference to the Geology and Zoology of the Floridian Peninsula.

A NARRATIVE OF RESEARCHES UNDERTAKEN UNDER THE AUSPICES
OF THE WAGNER FREE INSTITUTE OF SCIENCE
OF PHILADELPHIA,

BY

ANGELO HEILPRIN,

Professor of Geology in the Wagner Free Institute of Science; Professor of Invertebrate Paleontology at,
and Curator-in-Charge of, the Academy of Natural Sciences of Philadelphia;
Member of the American Philosophical Society, etc.
AUTHOR OF
"Contributions to the Tertiary Geology and Paleontology of the United States," "Town Geology:
The Lesson of the Philadelphia Rocks," and "The Geographical and Geological Distribution of Animals"
(International Scientific Series).

PUBLISHED BY THE
WAGNER FREE INSTITUTE OF SCIENCE
OF PHILADELPHIA.
1887.

INTRODUCTION.

THE Wagner Free Institute of Science was founded by the late William Wagner, a citizen of Philadelphia, who devoted a long lifetime to the study and advancement of the sciences, especially the different branches of natural history. Mr. Wagner, during his life, formed a large museum, a library and a collection of chemical and physical apparatus. He established annual courses of lectures on various scientific subjects, in which he personally took an active part, which were continued for thirty years, and which were always open free to the public. In 1855, under the above name, the Institute was incorporated by an act of the Legislature.

Mr. Wagner bequeathed his property to the Institute, vested in a Board of Trustees. Since his death in January, 1885, the Trustees have been actively engaged in carrying out his plans, and, in accordance with his views, have elected a faculty of four professors, to take charge of the museum and library, to give lectures free to the public, and to teach the method of, and also to make, research. The first annual course of free lectures was given by the faculty during the season of 1885 and 1886. The sphere of usefulness of the Institute will expand as the pecuniary circumstances are adjusted and will permit. That the benefits of the Institute shall not be restricted to its locality, but may be widespread as possible, the Trustees propose to make provision in aid of original research and the publication of its results, towards the increase and diffusion of knowledge among men.

Mr. Joseph Willcox, one of the Trustees, who had spent several successive winters in Florida, in speaking of his observations in that State, suggested the interest it would be to the Institute and to science to make an expedition to certain portions of the country, to make collections and investigations in their geology and fauna. Liberally offering his pecuniary and personal aid, and encouraged by the Academy of Natural Sciences, the Trustees of the Institute made the necessary provision, and the last winter sent Prof. Heilprin on the proposed expedition in company with Mr. Willcox. The results were valuable collections in zoology, and especially in geology, together with important investigations and discoveries in the latter, an account of which is presented in the following report by Prof. Heilprin. The well-observed facts of the report must greatly modify the opinions which have been generally held in regard to the geological construction of the peninsula of Florida; and altogether Prof. Heilprin's researches must be considered as an important contribution to science.

JOSEPH LEIDY,
President of the Faculty.

PHILADELPHIA, *January, 1887.*

PREFACE.

The following pages briefly narrate observations made, during the early part of 1886, in a region the greater part of which had most singularly escaped the attention of the scientific world. Although nearly seventy years have elapsed since the dominion of Florida was by act of Congress constituted into a territorial government, and upwards of forty years since admission into the Union was obtained, the State remains to the present day, as far as its geographical, zoological and geological features are concerned, very nearly the least-known portion of the national domain. So vague, indeed, has been the general scientific knowledge respecting the peninsula, that up to the time of our visit not even its broader geological aspects had been determined; that most fascinating of theories which ascribed the formation of this long stretch of country to the unceasing labors of the coral animal, and which, for nearly a full quarter of a century, received the almost undivided support of naturalists of both hemispheres, had only just begun to meet with its own disproof. The labors of a number of investigators in the northern part of the peninsula had already clearly demonstrated the inapplicability of the coral theory of growth to the facts presented in that section of the State, but we were as yet without data respecting the larger southern portion. With a view of adding to our knowledge more particularly of this region, a veritable *terra incognita* to science, the expedition, the details of which are here recorded, was, with the generous co-operation of the Academy of Natural Sciences of this city, and of Messrs. Joseph Willcox and Charles H. Brock, organized by the Wagner Free Institute of Science.

The personnel of this expedition consisted of the gentlemen above mentioned, of Captain Frank Strobhar, master of the schooner "Rambler," Moses Natteal, cook, and myself. Observations were conducted on the west coast as far south as the mouth of the Caloosahatchie, whence the expedition was deflected eastward into the Okeechobee wilderness. The general results of our geological investigations are summarized on pp. 65–67 of this report. The zoological researches were almost wholly confined to an examination of the littoral oceanic fauna, and to the fauna of the Okeechobee Lake region, which, I believe, had not hitherto been systematically investigated. Our facilities for work in this direction were, unfortunately, not quite as ample as could have been desired, and the results obtained perhaps not such as might have been anticipated. But the material collected, only a portion of which has thus far been elaborated, is sufficient to indicate the general faunal features of the region traveled over. The Gulf dredgings were all confined to shallow water, not exceeding twenty feet in depth.

A few words bearing upon the history of exploration of that mysterious body of water—Okeechobee—which had so long eluded research, and about which so many mythical fancies have clung, will not be amiss in this place.

It is not exactly easy to discover the earliest references to this lake. Captain Bernard Romans, who appears to have made an extended examination of the peninsula in the latter part of the last century, refers in his "Concise Natural History of East and West Florida" (p. 285)* to a large interior lake, unquestionably Okeechobee, as follows: "This is the river [St. Lucia], which, as i was told by a Spanish pilot of fishermen of good credit, proceeds from the lake Mayacco, a lake of seventy-five miles in circumference by his account. The man told me that he had formerly been taken by the savages, and by them carried a prisoner, in a canoe, by way of this river, to their settlements on the banks of the lake; he says, that at the disemboguing of the river, out of the lake, lies a small cedar island; he also told me that he saw the mouth of five or six rivers, but whether falling out of, or into, the lake, i could not learn of him; probably some of the many rivers i crossed in my journey across this peninsula, fall into it, and it is not improbable that St. John's river originates in it. The large river in Charlotte harbour [meaning the Caloosa, doubtless], by the direction of its course, meridian situation, and great width, i judge, might, perhaps, spring from the same fountain; however, the savages of Taloffo Ochasé told me, that in going far south, they go round a large water, emptying itself into the west sea, *i. e.*, Gulph of Mexico.

"Thus much have i been able to learn of this water, the exploring of which i always intended; whether there is really this lake, or not, i will not be positive, but the above circumstances, joined to a dark account, which the savages give of going up St. John's, and coming down another river, to go into some far southern region of East Florida (on which account the name of Ylacco, and the name given to St. Lucia by the savages, both conveying indecent meanings, are by them given to these rivers) seems to confirm it. That there is some such great water, is further to be gathered from the profusion of fresh water which this river,

---

* Printed in New York, 1776. From a pencil inscription in a copy of this work in the possession of the Academy of Natural Sciences of Philadelphia, it would appear that but very few numbers were ever distributed. It was sold by R. Aitken, Front Street, New York, "opposite the London Coffee House."

St. Lucia, pours down. Such is the immense quantity that the whole sound between the abovenamed island and the main, though an arm of the sea, situate in a very salt region, and in general two miles wide, is very often rendered totally fresh thereby; in so much, that it has made the very speculative Mr. De Brahm insist upon having seen mangrove stumps in fresh water. This lake has given rise to the intersected and mangled condition in which we see the peninsula exhibited in old maps."

It seems pretty certain from the above statement that little or nothing very definite was known of the lake before this period, except, perhaps, to a few who had accidentally visited its shores. The reference, however, to the "intersected and mangled condition" in which the peninsula appears in the earlier maps, clearly indicates that reports of the existence of such a lake had been broadly current, and not impossibly some accounts from personal observations had already been published. Indeed, on the map accompanying the "Account of the First Discovery and Natural History of Florida" of Roberts and Jefferys, published in London in 1763, the Laguna del Espiritu Santo is made to occupy approximately the position of our Okeechobee, although given a much greater extent than the lake actually occupies. A broad arm of the sea, designated the Bahia del Espiritu Santo, and corresponding in part with the modern Tampa Bay, is represented as opening into it from the west. Possibly the open water-way of the Manatee River suggested this connection. The lake is thus described (p. 18): "Laguna del Espiritu Santo is situated between the islands, extending from north to south about 27 leagues [81 miles], and is near eight leagues wide; it has several communications with the bays on the west side of the peninsula, as well as with the Gulf of Florida. The principal and best known entrance is about three leagues almost west from the Punta de Florida, which lies in 26 deg. 20 min. N. latitude. This entrance is two leagues nearly N. W., and at the end of it, in the lake, are two shoals and six islands, called the Cayos del Espiritu Santo; this large lake is as yet but little known." The entrance above referred to corresponds to a position a little to the north of Hillsborough River.

It is remarkable that these earliest accounts of the lake are but little less vague than those which have been published at various times during the succeeding hundred years, and surprising that our geographical knowledge of so large a portion of the national domain as is covered by the Okeechobee wilderness should have made such little headway. The great difficulty of gaining access to the region, doubtless, in great part accounts for this continued obscurity. Prior to the opening of the Okeechobee canal almost the only available approach was by way of the Kissimmee River. The beautiful waters of the Caloosahatchie, which are unquestionably fed by the Okeechobee swamps, lose themselves before the lake is reached, and thus what appears to be the direct water-way, was in reality, until the last two or three years, all but inaccessible. The difficulties of this passage are thus described by engineer J. L. Meigs, who, in 1879, undertook an exploration of the region under the direction of the Government: "On the 14th of March the united parties attempted to force a skiff, by wading, dragging and pushing, through the burnt stubble across the marsh intervening between Lakes Hikpochee and Okeechobee. After a day of exhausting toil, struggling through water and mire for the most part 2 feet deep, they arrived late in the afternoon within ¼ of a mile of the western shore of Lake Okeechobee, but their progress was arrested by vast beds of water-lilies, careless and frog weeds, and wild lettuce, filling the entire space between them and the lake, across which they were unable, by their united strength, to force the boat. . . . . Reluctantly the effort to enter Okeechobee was abandoned, and the parties retraced their steps, arriving in camp after midnight in a state of exhaustion after 16 hours of continued wading through water and mire" (Report of the Chief of Engineers, 1879, p. 865).

The author wishes to express his indebtedness to Mr. George W. Tryon, Jr., Conservator of the Conchological Section of the Academy of Natural Sciences, for much valuable aid received in the preparation of this report, and to the Levytype Photo-Engraving Co., of this city, for the very perfect rendering of the illustrations of new fossil species. The figures are reproductions direct from the specimens themselves.

A. H.

CONTENTS.

THE WEST COAST OF FLORIDA.

Configuration of the coast line, page 1; Vegetation, 2; Homosassa River, 3; Cheeshowiska River, 4; Nummulitic deposits, 6; Animal life, 7; Anclote Keys, 7; Tampa and Hillsboro Bays, 9; Ballast Point, 10; Manatee River, 12; Sarasota Bay, 14; Fossil man, 15; Little Sarasota Bay, 16; Gasparilla Inlet, 18; Charlotte Harbor, 21.

THE CALOOSAHATCHIE.

Characters of the stream, 22; Natural history of the, 24; Geological features, 26; Fort Myers, 26; Thorpe's, 28; "Floridian" formation, 29; Fort Thompson Limestone, 32.

THE OKEECHOOBEE WILDERNESS.

Okeechobee canal, 34; Alligators, 35; Lake Hikpochee, 36; Everglades, 37; Fauna of the canal, 38.

LAKE OKEECHOBEE.

Geography of the lake, 39; Physical features, 41; Observation Island, 43; Taylor's Creek, 44; Vegetation, 45; Animal life, 46; Alligators, 47; Serpents, 48; Fauna of the lake, 50.

GEOLOGICAL OBSERVATIONS.

Historical survey, 52; Outline of formations, 56; Coral theory of growth, 56; Geology of the Homosassa (Oligocene), 3, 57; of the Cheeshowiska (Nummulitic), 6, 57; of the Pithlachascootie (Oligocene ?), 60; Ballast Point, Hillsboro Bay (Miocene), 10, 61; Hillsboro River, 12, 62; Manatee River (Miocene), 13, 63; Sarasota Bay, 14, 64; Little Saratosa Bay (Mio-Pliocene?), 17, 64A; Caloosahatchie (Pliocene), 28-32, 64B; General summary and conclusions, 65.

PALEONTOLOGY OF THE PENINSULA.

Fossils of the Pliocene deposits ("Floridian") of the Caloosahatchie, 68-104; Miocene fossils of Ballast Point, Hillsboro Bay, 105-123; Fossils from localities north of Ballast Point, 124; Table of Atlantic and Gulf Tertiaries of the United States, 127.

THE WEST COAST OF FLORIDA.

Our observations on the west coast of the peninsula were confined to the tract included between Cedar Keys and the mouth of the Caloosahatchie (Punta Rassa), or over an area measured by somewhat less than three degrees of latitude. Along this entire reach the coast is very low, rarely rising more than from five to ten feet above water-level in the immediate neighborhood of the ocean border. The most elevated point would seem to be the dune at Clearwater, which, according to an official railroad survey, rises to a height of thirty-two feet; a portion of this "bluff" is made up of the remains of an ancient Indian shell mound, the wreck of which is clearly indicated in the large conchs, *Fulgur perversum*, *Melongena corona*, etc., which lie scattered about. Immediately back of the town of Tampa, about a quarter of a mile up the Hillsboro River, and a little to the inland of the left bank, the solid rock rises to a height of some fifteen or twenty feet, but southward, again, even these minor elevations disappear, and the coast for the greater distance presents the appearance of a tide-level reach.

Contrary to what is generally supposed, solid rock enters very largely into the formation of the peninsular border, and its outcrops can be observed as well without as within the river channels. Thus, it is exposed on the Homosassa River a short distance (a mile or more) above its mouth, at various points on the Cheeshowiska, on John's Island at the mouth of that river, along the Pithlachascootie, on Clearwater beach, at Ballast Point on Hillsboro Bay, at the locality above Tampa already indicated, and at numerous other points. The rock is almost everywhere a more or less compact limestone, heavily charged with fossil remains, and at a few localities, as at Ballast Point and along the bed of the Hillsboro River, largely impregnated with silica, forming a tough siliceous matrix which readily yields to the hammer. Where the solid rock is not visible the eye rests upon a beach of homogeneous white or yellowish sand, which in some places is almost wholly deficient in shell-fragments, while in others it is literally packed with them. The most extensive shell-bank observed forms the ocean front between Little Sarasota Inlet and Casey's Pass, where, in a thickness of 4-5 feet, the greater number of the molluscan species now inhabiting this part of the coast can be found. A true coquina rock was found at the entrance to Little Sarasota Inlet, and again on Philippi's Creek (tributary of Big Sarasota Bay)—at the latter locality overlying a fossiliferous arenaceous limestone—the first time, I believe, that it had been noticed on the west coast of the State.

Everywhere along the border the oceanic floor shelves very gradually, so that at even considerable distances out to sea only a few feet of water can be obtained. Whether or not a distinct channel depression exists beyond the mouths of all the various streams discharging on the coast, our means did not permit us to determine with any amount of positiveness, but it would seem that such is the case in at least some instances. Admitting this configuration of the bottom, it could readily be accounted for on the hypothesis of a steady or continued subsidence of the land, to which numerous facts, not necessary to be indicated in this place, seem to point. It is barely possible that the acidulated waters of the outflowing streams could have produced any measurable amount of subaqueous erosion. The mouths of the northern streams more especially—Homosassa, Cheeshowiska, etc.—are very largely obstructed by oyster-reefs, which, in some places, appear above water-level during low water, and render difficult a passage of the channel to all but the smallest craft. These reefs are rapidly developing, and must ultimately completely bar the passages.

The vegetation along the west coast may be said to be fairly luxuriant. A semi-tropical character prevails in the northern tracts, especially well-marked along the upper Cheeshowiska, where the forest unfolds itself in its noblest and most magnificent proportions. The bay, water-oak, live-oak, cypress, and palmetto stand out as the most prominent features of this confused vegetable maze, whose penetrability is rendered possible only through the small bayous or narrow water-courses which partially enter the inner recesses of the wilderness. Southward, as at Dunedin, etc., where the thickness of the sand deposit very materially increases, the virgin forest largely disappears, and is replaced by a much weaker growth of yellow-pine and saw-palmetto, the latter forming an undergrowth rarely rising above three or four feet. This stretch of pine land extends for a very considerable distance down the shore, relieved here and there by recurrences of the more vigorous tropical jungle, a feature observed along some of the larger water-courses. Much of the thicket has been removed from the banks of the Hillsboro, but on the Big Manatee, a short distance above Braidentown, the palm forest assumes its pristine character. Along the protected bays and lagoons, formed by the outlying sand keys—Sarasota Bay, Gasparilla—the outer border more especially (or the keys) is fringed by a dense growth of mangrove, which continues with but slight interruption to the southern end of Charlotte Harbor. Its greatest development is seen here, where the "bushes" attain the dimensions of small forest trees. At the time of our visit the foliage wore an autumnal aspect, the sere and purple leaves, the result of the recent cold wave, severely recalling the end of a northern season.

OFF ST. MARTIN'S REEF AND THE HOMOSASSA RIVER.—Somewhat to the north of the mouth of the Homosassa River the coast is bordered by a long line of broken reef, under whose lee we anchored the first night after leaving Cedar Keys (February 14). Here, in water of 6–7 feet depth, we obtained, by means of the scoop-net and hook, numerous sponges and several corals, the former of which thrives here in abundance. A large specimen of the logger-head sponge was found to measure nearly 17 inches in greatest diameter and eight inches in height. A number of these were immediately put in alcohol, and others placed on deck to dry, but the highly offensive odor resulting from decomposition necessitated their early restoration to the oceanic medium. The corals belonged to the genus Orbicella (*O.* [*Siderina*] *galaxia*), and were dead, but traces of the animal substance, still highly colored, showed that their existence had but recently ended. I believe this is the most northern point in the Gulf at which coral life has thus far been determined.

On the following morning we pushed our skiff up the Homosassa. My own observations were restricted to the lower two miles, but special information as to the upper course was brought to me by Mr. Willcox, who on many previous occasions ascended the stream to its source, for a further distance of about six miles. The fountain is described as a transparent pool of considerable depth, lodged in a basin of compact limestone, probably of the same character and age as that which appears not very far from the mouth of the river. At Wheeler's (left bank), somewhat more than a mile from the Gulf, this limestone was exposed at the time of our visit some one and a-half to two feet above the water, which has honeycombed it in all directions. Great numbers of *Mytilus hamatus* are here attached to the rock. A number of caverns and sinks appear some little distance from the bank, evidently excavated by the water of the stream gaining access into the numerous fissures that traverse, and are being cut into, the fundament. Large lumps of rock, collected from a well-digging, show an unmistakable fossiliferous character, but the fossils are mainly in the form of casts or impressions, and barely permit even of the determination of genera (mollusks). The immediate border rock is much more compact, and in a rapid inspection might be taken to be non-fossiliferous, but a magnifying lens readily reveals its true nature. The innumerable casts and impressions of the niliolite genera Biloculina, Triloculina, Quinqueloculina, Sphæroidina, and other kindred forms, clearly betray its foraminiferal structure, and point to its deep-sea origin or formation. I propose to designate this limestone the "Miliolite Limestone," representing one of four distinct types of foraminiferal rock found in the State, the others being the Orbitoide, Nummulitic, and Orbitolite limestones. It certainly forms part either of the Oligocene formation or of the Upper Eocene, much more likely the former, and may possibly be in part synchronous with a portion of the West European miliolite rock.

The banks of the Homosassa are beautifully wooded, presenting in the profusion of the palmetto and yucca growths a partially tropical aspect. I was much surprised at the general absence of indications of animal life, the forest being as silent as the inner recesses of our more sombre northern wilds. An occasional flock of herons or white ibises would, perhaps, for a moment cloud the firmament, a mullet spring from the water, or the cardinal grosbeak peal its clear whistle, but otherwise an impressive silence pervaded the entire solitude. This was in marked contrast to some of our later experiences, and was probably accidental. A few days before our arrival, I was informed by Mr. Wheeler, a large spotted cat, of a light color, and somewhat smaller in size than the panther or Florida lion, had been killed in this neighborhood. To my numerous inquiries as to its identity with a species of lynx or catamount, or one of the better-known tiger-cats, I was only able to elicit a negative reply. Its distinctness from all of these forms was independently confirmed by our cook, an experienced huntsman of the upper Homosassa, whom we obtained later in the course of the day. Is it possible that the ocelot is an inhabitant of these wilds, and that it should have escaped the notice of traveling naturalists?

OFF THE CHEESHOWISKA RIVER.—On the 16th, and 17th the "Rambler" remained at anchor off the mouth of the Cheeshowiska River, giving us an opportunity to explore several miles of this exquisitely beautiful stream. The vast oyster reef at its mouth rendered the passage of the channel intricate, and in a manner dangerous, and on our return journey one of the boats had to be partially relieved of its load and hauled over the shells. On John's Island, which guards the mouth of the river within the reef, we found innumerable aboriginal implements, some very rude, others more perfect, fashioned from a siliceous rock which appears to be identical with the rock exposed on a small island about three miles S. E. of the mouth of the Homosassa. The great number of partially finished implements and chips, and the masses of the nearly crude rock lying about, leave little doubt in my mind that the island, as first suggested by Mr. Willcox, was the true factory where these implements were manufactured.

At low water a somewhat spongy limestone, containing numerous molluscan impressions and a few Orbitoides, appears on the ocean front;

in nearly all instances where large masses of this limestone are examined they are seen to be bordered by, or enclosed in, a limestone of a somewhat different character and color, which, in addition to numerous fragments of marine-type fossils, contains the remains, beautifully preserved in many cases, of freshwater organisms, such as Vivipara (*V. Waltonii*) and Ampullaria (*A. depressa*), and of species which still inhabit the existing waters. The working over or re-formation of the original limestone is thus established beyond a doubt. The same limestone is exposed about a mile further up the river, where a clump of palmettos marks a turn in the stream.

For some distance above and below this point the region may be described as a grass or meadow land, subject to periodical overflows from the numerous tidal channels that intersect it in all directions, and which in a measure disguise the principal stream. Terra firma appears only at intervals, but is then clearly marked by the inevitable growth of palmetto which clothes it. The tall sedges were alive with the busy and ever garrulous grackle or "jackdaw" (*Quiscalus major*), whose familiar notes were poured forth in one almost continuous strain. We observed numerous egrets and snow-herons, and an occasional blue-heron. Where perching room was afforded we were almost sure to meet with one or more individuals of the snake-bird (*Plotus anhinga*), with expanded drying wings, or the dreamy cormorant quietly watching its opportunity. Two raccoons appeared on a mud flat within easy gun-shot of our boat—remarkably enough, if we except a limited number of deer, rabbits and squirrels from the upper Caloosahatchie—the only terrestrial mammals encountered by our party during the entire trip of six weeks.

At about four miles above its mouth the stream emerges from the virgin forest, which extends in an almost unbroken belt to the limits of vision. To one who has never before contemplated the beauties of a southern vegetation it is impossible to convey an idea of the magnificence of this semi-tropical jungle—the endless variety of contrasts that are presented in the vegetable outlines, the luxuriance that is ever manifest, and above all the brilliant greens that peer refreshingly through the outer dense masses of foliage! The eye never tires of following the delicate tracery of the innumerable climbing plants that hang festooned from the arms of some noble forester, or shroud the palmetto in a garden of its own, or of gazing upon the rugged trunks of the live-oak and water-oak that rise above these, and rear their crowns, heavily draped in Spanish moss, against a firmament of deepest blue. Everything was bright and fresh, and it seemed as though a region had been found where neither the chilling blasts of winter nor the parched tongue of summer had as yet been able to penetrate. I observed a marked deficiency of plants in bloom; indeed, as far as my own observations went, all the visible

flowers were confined to a limited number of water or marsh-plants—
lily, flag, etc.

At Loenecker's, a short piece beyond the outer border of the woods,
is the locality whence Mr. Willcox obtained the nummulites described
by me some four years ago as *Nummulites Willcoxi*, the first representatives of the genus that had up to that time been found on the North
American continent. The exact spot is a ploughed field, cleared from
the bush, about five minutes from the right bank of the river, and
elevated, according to a rough estimate made by us, about 4–6 feet above
the surface of the water. The rocks containing the fossils occur loose
in the soil, and, doubtless, have in great part been thrown up by the
plough. No trace of a solid outcrop was anywhere visible. While,
therefore, the presumptive evidence is that these rocks have been moved,
and are, consequently, no longer in their normal positions, yet it is highly
probable that the parent rock is not far distant. Indeed, I am assured
by Mr. Willcox that he has observed the "Nummulitic" *in situ* at a
locality distant some fifteen miles in a N. E. direction. We found the
rock at Loenecker's literally charged with the tests of nummulites and
orbitoides, so much so, in fact, as to present the appearance of being
built up almost entirely of the hard parts of this lowest group of
animal organisms. Many of the larger fragments or boulders, as on
John's Island, were encased in a newer matrix of considerably darker
color, in which the remains of the recent shells already referred to, and
some others—Mytilus ?—were found embedded in a beautiful state of
preservation. To what extent this newer freshwater formation extends,
or if it constitutes but a mere strip formed as a fringe to the older
(marine) deposits, our means did not permit us to determine. But
manifestly, there must have been considerable changes in the topography
of the region since the river limestones, of comparatively recent date,
were added to, and united with, the marine limestones of the Nummulitic (Oligocene) period.

In company with one of our boatmen I ascended the river for a
further distance of about a mile and a-half, in the hope of discovering an
outcrop. This we found in a mass of rock jutting out from the bottom
of the channel, but barely reaching the surface of the water, and in a
number of rounded ledges whose outlines we could distinguish through
the limpid waters. With the assistance of a mattock we succeeded in
detaching several fragments, but the toughness of the rock, and the
difficulty of striking below the water, prevented us from obtaining as
many specimens as we should have desired. Much to my surprise
the rock contained not a fragment of either of the two forms of Foraminifera which were so abundant at Loenecker's, and so eminently serve
to characterize the formation in which they occur. Indeed, the only

fossil impressions were those of two species of bivalves, which, from their imperfect state of preservation, can only doubtfully be referred to Modiola and Cytherea.

The lateness of the hour prevented any further exploration in this direction, and compelled us to retrace our steps in the direction of the schooner. Evening had now fairly set in, and the exuberance of animal life that everywhere greeted us on our ascent vanished as if forever. A stray flock of herons or ibises might still be seen wending its path in the direction of some secluded heronry, an occasional hawk gracefully circling in its aerial height, but the hushed silence of eventide hung like a pall over the landscape. The numerous turkey-buzzards which earlier in the day hovered like so many spectres over the objects of their special adoration, flitting their shadows, as ethereal clouds, across the emerald wall of the forest, now clung noiselessly to the withered branches of some former pride of the wilderness. Thirty or forty, or even more, of these birds could frequently be counted on a single tree, perched like so many black statues in silent contemplation of the visions of departed day.

Only the waters still gave evidence of unabated animal vitality. The myriads of fish—mullet, skip-jack, etc.—that disported in the tangle of grass which everywhere covered the floor of the river, formed a most interesting picture, and one decidedly refreshing in its novelty. We observed two individuals of the alligator-gar.

ANCLOTE KEYS.—In our anxiety to make the best of our sailing time we grounded on a grass shoal just beyond Anclote, and anchored for the night (18th). Low water early in the morning permitted of a considerable amount of wading, and we had thus a very favorable opportunity presented for studying the zoological features of our anchorage. We found a spinose star-fish (Echinaster sp.?) fairly abundant, and secured a number of specimens, but this was the only species of the group observed here. There were no urchins—at least, we failed to detect any if present. We hooked up a number of the bright yellow sponges of the genus Rhaphyrus (*R. Griffithsi*), and with our landing-net scooped in a fair supply of one or more species of simple ascidians (*Ascidia ovalis?*). Much of the grass was found coated with the compound masses of a species of Botryllus. Among the other forms of animal life taken here were the scallop (*Pecten nucleus*), sea-spider (*Libinia canaliculata*), cow-fish (*Ostracion quadricorne*), zebra-fish (*Chilomycterus geometricus*), and a pipe-fish (Syngnathus), besides a considerable number of diminutive molluscan forms (Columbella, Nassa, Cerithium, etc.). The tulip-shell (*Fasciolaria tulipa*) was fairly abundant.

We floated off on high-water, and steered southward to Dunedin.

Here we were informed of the recent finding of mastodon remains at a locality some three miles distant, but, unfortunately, the limited time at our command did not permit us to visit the spot. I followed the coast line for about a mile south of the town, through a dreary sand tract of yellow pine and saw-palmetto, in the hopes of finding an outcrop of rock, but without success. At Clearwater the most elevated point of land on the west coast, the dune, to which reference has already been made in the introduction, makes its appearance, rising to the very modest height of 32 feet. Immediately north of the landing a tough siliceo-calcareous rock juts out from the ocean beach, but the scanty and unsatisfactory condition of its contained organic remains precluded the positive determination of its position in the geological scale, although in all probability a belonging of the Oligocene series.

We grounded again just off the passage of Sand Key, and remained becalmed and anchored for a full day and a half in a somewhat unprofitable position. The sand-beach on the ocean side of the key was literally packed with the shells of *Venus cancellata*, but I failed to observe a single live animal of that species, although undoubtedly an inhabitant of the adjoining waters. Among the living Mollusca, *Strombus pugilis*, *Fasciolaria tulipa*, and *Fulgur perversum* were sufficiently abundant; the last, however, was most numerously represented on the inner side of the key, on the mud-flats, where its egg-capsules or spawn-ribbons, many of them evidently only recently deposited, lay scattered about. Although our attention was centred in that direction, we observed but few individuals in the virtual act of depositing this ribbon, but, doubtless, many others performing the operation escaped our notice. In these instances the animal was in greater part buried in the sand or mud, the spawn-ribbon being anchored by the smaller end to a shell or pebble. On lifting the animal from its cover, the ribbon was almost immediately ejected.

The crown-conch (*Melongena corona*), judging from its abundance, evidently found here a most congenial home among the mud-flats. In shoal water the bottom was covered for acres with two species of sea-anemone (*Cerianthus*), whose habits could be very easily studied through the transparent water. Owing to the depth, a foot or more, to which the animals were immersed in the sand, and the tenacity with which they held on to their anchorages, it was almost impossible to obtain perfect specimens. I observed that the body portion, or external tunic of the animal, was much more sensitive to impacts than the tentacular; thus, if touched on the body, the animal almost instantly withdrew, whereas if the tentacular portion only was touched, there was frequently a decided hesitancy on the part of the animal to withdraw. This was not always the case, however. When once retracted the animal remained in this condition for a considerable period. I noticed, too, that a cloud of sand or mud precipitated over the animal produced no sensible effect upon its movements. Pelagic forms of life, such as jelly-fishes, were decidedly scanty, and it must be admitted that their absence was a source of no little disappointment. Unfortunately, we were not sufficiently equipped for prosecuting zoological researches by night, otherwise as far as the pelagic fauna is concerned, our efforts at collecting might have been attended with better success. Toward evening we obtained a number of Idyas of the form of *Idya roseola*, only colorless, which were retained alive in a basin of sea-water for very nearly two days.

While in our enforced captivity off Sand Key we were much interested in watching the habits of the hundreds of pelicans, cormorants, and gulls that frequented a small sand island or shoal in the middle of the harbor. The pelicans and cormorants seemed to mingle indiscriminately into a single household, but the gulls evidently preferred an independent position of their own, ranging themselves in linear series, lumps of silvery white, like so many sentinels to a flock.

TAMPA AND HILLSBORO BAYS.—We left our anchorage early on the 21st, and the same evening made Point Pinellas, at the entrance to Old Tampa Bay. The passage of Boca Ceiga (John's Pass) was effected without much difficulty, although its direction had to be made from the mast-head. Recourse to the mast-head has frequently to be had in the navigation of Florida waters, owing to the numerous shoals that bar the passages, and the difficulty of their determination from a low level. Even the most experienced pilot will consider himself fortunate if he escapes one or two trials of stranding during a day's journey, and there are probably very few who can claim immunity from the results of what the non-initiated might consider bad sailing.

We dragged in shallow water just after passing the Boca Ceiga, but the dredge brought up little of consequence; the haul consisted almost exclusively of myriads of *Venus cancellata* and *Nassa trivittata*. At our anchorage inside of Point Pinellas we secured a specimen of a beautiful rose Aurelia, measuring some seven inches across the disk, the first of our jelly-fish captures; a species of brittle star, *Ophiolepis elegans*, was very abundant, and several individuals could almost invariably be obtained from every bunch of grass that was scooped up by the net. We tried the experiment of night collecting, and obtained a number of forms that would otherwise probably have been lost. The young of an undetermined species of fish, and numerous small crustaceans were especially attracted by the glare of our lamp, and through it we also obtained a specimen of the balloon-fish (*Tetrodon turgidus*), and a half-beak (*Hemirhamphus unifasciatus*), which, in its eager survey of the artificial "moon," skipped over it and landed in our boat.

The shore at this point was strewn with dead fish, more especially with the remains of the cavalle and cow-fish, an index of the disastrous effects of the cold wave that had recently swept over the greater part of the State. It was almost inconceivable that a sudden lowering of the temperature could have had such a marked effect upon the vitality of animals inhabiting the sea, but the proof of such effect was everywhere apparent, and could not be argued round by any amount of logical theorizing. The worst effects were, however, to be noted further down the coast.

At about noon of the next day we made Ballast Point, four-and-a-half miles southwest of Tampa, a spot made famous to geologists and mineralogists through its numerous silicified shell remains, retained in the most exquisite state of preservation, and the coral-chalcedonies that occur in the form of organic geodes. In the yellow limestone that makes the basal outcrop at this locality I immediately recognized the foraminifer which Conrad some forty years previously had described as *Assilina* (*Nummulites*) *Floridana*, and from which the age of this portion of the peninsula had been considered established. Conrad had evidently entirely misinterpreted the nature of his fossil, inasmuch as his drawing represents an imperfect individual, or one in which through an irregular removal of the shell layers, exposing a gradational elevation of the disk, the involution of the whorls is made to assume the form of a spiral, instead of that of a series of concentric rings. The rock here was crowded with the disks of this foraminifer—many of them in the condition figured by Conrad, others perfect—which, as I had already suspected, is no nummulite at all, but a member of the very different genus Orbitolites. This is the first record of this somewhat rare genus being found on the North American continent. Among the other fossil impressions I detected those of *Venus penita* and *V. Floridana*, also described by Conrad, and of a number of generic types the specific characters of which were too much obscured to permit of clear definition.

Numerous angular boulders of a tough siliceo-calcareous blue rock, also densely charged with fossils, rest on the yellow limestone above mentioned, but the relative sequence of the two formations could not be determined at this point. Several of the fossil species occurring in this rock appeared also to be contained in the limestone, but the former was distinguished from the latter by the total absence of the foraminifer Orbitolites and by the presence of vast numbers of casts and impressions of a species of Cerithium. This genus, one of the most abundantly represented and distinctive genera of the Eocene, Oligocene and Miocene formations of Western and Central Europe—indeed, of nearly all regions where the early and middle Tertiary deposits are developed to any extent—had hitherto been known only by stray individuals in this country, and its absence constituted one of the negative faunal features by which the American Tertiary formations were distinguished from the European. The discovery of a true Cerithium rock, therefore, becomes an interesting feature in connection with the geology of this region.

The country about here presents the appearance of an inhospitable sand tract, thinly dotted with pine groves, and covered with a low growth of saw-palmetto (*Sabal serrulata*), the reputed home of the rattlesnake and moccasin. We found a species of prickly-pear (*Opuntia*) in bloom. A short piece above Ballast Point proper, at Newman's Landing, is the outcrop which furnishes the silicified shells and chalcedonized corals to which reference has already been made. Unfortunately, the position of the outcrop is such as not to permit of an absolute correlation with the deposits exposed at the Point, but I feel satisfied that it cannot represent an age very different from that of the yellow limestone, with which it holds several molluscan species in common. Its position is in the Lower Miocene series. The greater number of the species are here imbedded in a marly matrix, from which they can be readily removed by means of a pick or mattock. With few exceptions all the forms are extinct; a limited number of them are found in the nearly equivalent deposits of the island of Santo Domingo. The corals are principally astræas and madrepores, but of a number of distinct species; as far as could be determined they form a border fringe, the remains possibly of an ancient reef. What led to their hollowing out in the form of geodes, and the manner of the substitution of chalcedony for the carbonate of lime, are problems still awaiting solution; doubtless, heated waters, largely impregnated with silica, were directly involved in the operation, but just why the outer layers of the coral masses should have been preserved, while the inner parts so readily yielded to solution, is not exactly apparent.

The day after our arrival in Tampa, I, in company with our cook, made an examination of the lower Hillsboro, sailing up the river in our skiff for a distance of about five miles. The shores were almost everywhere very low, rarely rising more than five or six, or a dozen, feet above the water, except immediately above the town, where, a short distance from the left bank, there is a somewhat abrupt rise of possibly twenty feet or more. A fairly luxuriant growth of woodland covers both banks for the greater distance, but we found few traces of that primeval forest which at one time, doubtless, graced this region as it still does the region of the Cheeshowiska. Nor did the forest present here the same tropical appearance which it unfolds in the region further to the north; the bay and water-oak still continue as some of its dominating features, but there is a very noticeable deficiency of palmettos, and, in their stead, a marked increase of the coniferous element—yellow pine and swamp cypress.

Not much more than a quarter of a mile above Tampa, and just below the ship-yard, a tough siliceo-calcareous rock, identical with that found at Ballast Point, appears on the left bank immediately on water-level; the same rock is visible on the right bank at a further distance of about a half mile, and reappears again at intervals of three, four and five miles. There can be no doubt that it forms the bed of the stream for this distance. It can be readily identified by its numerous Cerithium remains, the same as we found impressed in the rock at Ballast Point. At Magbey's Spring, a short piece above the ship-yard, we found water issuing from a yellow and white limestone, containing numerous fossils; large sink-holes expose the fossiliferous limestone, crowded with shell remains and the *Orbitolites Floridana*, for an extent of some ten feet. Owing to the very limited nature of the exposure I was unable to determine its true dip, but as the locality is distant not more than a few hundred feet from the river, and rises above it some fifteen or twenty feet, there can be no doubt that the rock in question overlies that which appears immediately on the river front, and which, as has already been said, almost positively forms its bed for a distance of several miles. In this section, therefore, we have established the relation existing between the two rocks exposed at Ballast Point. The locality at Magbey's Spring is the only one on the Hillsboro River where we observed the Orbitolite limestone.

During the day's journey my attention was called to an individual of the Florida "mud-puppy," but I was unable to approach the animal sufficiently near to determine whether it was a Necturus or not. Turtles were surprisingly abundant, and their splash, when dropping from an overhanging bough, could be heard at frequent intervals around the turns of the stream. Nine individuals, of possibly more than one species, were seen on a single raft, sunning themselves in pleasant ignorance of impending danger. I much regretted not being able to visit the falls of the Hillsboro, about three miles beyond the furthest point reached in our exploration, where the ledge of rock over which the water is precipitated is said to be largely coralliferous, and of the same character as that observed at Ballast Point.

MANATEE RIVER.—We left Tampa toward evening, pushing off with falling tide, and headed for the Manatee River. The dredge was thrown over in the mouth of that stream, and struck on an Anomia bank. The dead shells of *Venus cancellata* were brought up in great quantity, together with a number of crabs, a species of Lima, and several individuals of the common sea-urchin of this part of the coast, *Toxopneustes variegatus;* depth of water about 12 feet.

It was our intention to explore some of the islands in Terraccia

Bay, where fossil remains were reported to be abundant, but at Braidentown we were informed that a fossiliferous exposure was presented a few miles (5-6) above the town at a locality known as Rocky Bluff, and we accordingly determined to visit that spot. The "bluff" we found to be a ledge of rock, rising about two or three feet above water-level at the time of our visit, and consisting of at least two well-defined layers—a basal white "marl" and yellowish sandstone, and an overlying siliceous conglomerate. The latter is almost entirely deficient in organic remains, whereas the marl is densely charged with them. Among the recognizable forms occurring here I determined a number of well-known and distinctive Miocene species of mollusks, such as *Pecten Jeffersonius, P. Madisonius, Perna maxillata, Venus alveata, Arca incongrua*, etc., which left no doubt as to the age of the deposits in which they were imbedded. The existence of a Miocene formation in this portion of the peninsula was entirely unlooked for, and its discovery, therefore, the more significant and interesting. A further exploration of this bed was made on the succeeding day, but without adding much that was new to our stock of information obtained the day previous. The white bed thinned out and disappeared after a short distance, but the yellow sand-rock, largely honeycombed, and containing much fewer fossils, many of them identical with the forms of the marl, continued up the river to the furthest point reached by us. I observed and collected many fragments of manatee bones, ribs principally, but am not prepared to say that any of these were of a fossil character, although their position might have led one to suppose that they had been washed from the bank. Mr. Willcox, however, assures me that he observed several pieces concerning the fossil nature of which there could be no doubt.

In the hope of discovering a more extended outcrop in the interior, and of securing a position whence a general survey of the region could be obtained, I attempted to penetrate the dense growth of palmetto that here descends to the river's bank, but owing to the obstruction presented by the large fan-leaves, and the difficulty of determining landmarks in a tract where the component vegetable elements so greatly resembled one another, was compelled to desist after wandering about three-quarters of a mile. The forest is here evidently largely of second growth, but few of the trees, mainly palmettos, attaining to more than mediocre proportions. Mr. Brock secured two alligators before leaving the river, the larger of which measured about nine feet in length. About a mile above the point where we made our geological examination the river-bank was packed with the remains of dead fish, which were lying heaped up in windrows of tens of thousands of individuals. No such wholesale destruction of the shore-fishes appears to have been known to any of the inhabitants.

SARASOTA BAY.—We were informed that at Hunter's Point, near the northern end of the Bay, we would find a coral rock or formation skirting the shore; I was naturally anxious to determine the accuracy of the statement, inasmuch as no reef formation had been reported from the region so far to the north. The rock in question turned out to be a vast mass of growing Vermetus (*V. varians*), which from a short distance actually presented the appearance of a clump of rocks. A limestone of an analogous structure crops out in the meadow a few hundred feet from the shore. The same growth of Vermetus reappears at Whittaker's, a few miles further down the bay, where the matted tubes of the gasteropod form organic "boulders" or reefs stretching over acres of territory, one of the most striking features of this part of the coast. A yellow sand-rock, some three or four feet in thickness, appears at this point on the shore margin; its general aspect bears the impress of a recent formation, but I found in it the casts of one or more species of coral of a facies new to me, which, in the absence of other definable organic remains, led me to suspend judgment as to the age of the deposit. The same coral I afterwards identified in a more compact, and much more fossiliferous, limestone occurring on White Beach, Little Sarasota Bay.

On Perico Island, where we landed for the purpose of skinning our alligators, we found vast numbers of the common fiddler-crab of the coast (*Gelasimus pugilator*), which, in apparent concerted action, were hurrying from the sea-border into the interior, passing far beyond the line of their burrows. So numerous were the migrating hordes, that in many places they literally obscured the beach, and the noise of their progression was like that produced by a wind moving a heavy accumulation of autumn leaves. The border of the island was covered with a heavy fringe of mangrove, on whose aerial roots, considerably above water-level, we found the parasitic oyster (*Ostrea parasitica*) clinging in great abundance. The interior of the island supports a stunted growth of saw-palmetto, and the usually accompanying yellow-pine. We found a moccasin coiled on the leaf-stalk of a palmetto, about two feet above the ground—the first ophidian met with on our trip; the animal, although plainly cognizant of our approach, made no attempt to attack, and but a very feeble one to escape, and was consequently secured without much difficulty.

At a locality known as Mrs. Hanson's, opposite to which we anchored for the night, I was conducted to a spot where it had been reported a human skeleton lay embedded in the rock. My misgivings as to such a find were naturally very great, but I could not resist the temptation of satisfying myself personally in the matter, even at the risk of appearing over-credulous to my fellow-companions. The rock I found to be a partially indurated ferruginous sandstone, removed but a short distance from the sea, and but barely elevated above it; the condition of its exposure was, doubtless, the result of recent sea-wash. I was much surprised to find actually embedded in this rock, and more or less firmly united with it, the skeletal remains of a mammalian, which I had little difficulty in determining to be the genus Homo. Most of the parts, including the entire head, had at various times been removed by the curiosity-seekers of the neighborhood, but enough remained to indicate the position occupied by the body in the matrix. The depression which received the head was still very plainly marked, but unfortunately the outline had been too much disturbed to permit of any satisfactory impression being taken from it. I was able to disengage from a confused mass of stone and skeleton two of the vertebræ, which Dr. Leidy has kindly determined for me to be in all probability the last dorsal and first lumbar. The distinctive cancellated structure of bone is still plainly visible, but the bone itself has been completely replaced by limonite.

How great an antiquity these human remains of iron indicate, I am not prepared to say. That they are very ancient there can be no question, considering the nature of their fossilization, and the position which they occupy; but to which exact horizon in the geological scale they are to be referred, still remains an open question. I in vain searched the region for geological landmarks by which the special bed containing the remains could be correlated, but in vain. I could find no trace of any other fossil in the deposit, nor, owing to the low position of the bed, and the absence of overlying deposits of any magnitude, could its homotaxis with reference to the fossiliferous deposits occurring elsewhere on the bay be ascertained. The probability naturally lies with the Post-Pliocene age of the deposit, but for aught we know to the contrary, the age represented might in fact be Tertiary. At all events, as has already been stated, the remains are very ancient, and not impossibly they represent a period as far (if not further) removed from the present one as is indicated by any other human remains that have thus far been discovered.

About three-quarters of a mile below Mrs. Hanson's a compact terrestrial sand-rock, containing numerous individuals of several common forms of recent snail (*Polygyra volvoxis*, etc.), and evidently representing a modern formation, is exposed at water-level, extending for some little distance up the channels that have been left by the retreating waters. The presence of this hard rock of terrestrial origin on the immediate ocean front, and in the very path of existing waters, coupled with the circumstance of the complete absence of associated marine forms of life, renders it more than probable that this portion of the coast has quite recently been undergoing subsidence. It is true that the encroaches of the sea might be attributed to a simple washing away of the coast line, but this hardly appears probable in view of the resisting nature of the rock, and the fact that it rests horizontally and shelves for some distance, at least, under the sea.

From this point Captain Strobhar and I made a diversion in favor of Philippi's Creek, a tributary of the bay. We found plenty of water in the stream itself, but the approaches to it, owing to the widening out of the channel, were very shallow, and for a considerable distance our skiff had to be dragged over the bottom. The difficulties of the passage were further increased by the numerous islands, largely overgrown with mangrove, which interpose themselves in the mouth of the creek, rendering the channel very intricate. Almost at the mouth of the stream, and at several points above the mouth, we found a true compact coquina rock, some three to four feet in thickness, the first time, I believe, that such a rock had been noted to occur on the west coast of the peninsula. The shell fragments composing it were largely triturated, and in most instances not even the genera of mollusks represented by them could be identified. Underneath this rock, where present, there crops out a yellow arenaceous limestone, which is exposed at various points along the stream, rising about two feet above water-level. It contains coral impressions and numerous shells, many of the latter apparently identical with forms found in the yellow rock of the Manatee River (*Pecten Jeffersonius*, etc.), and representing either a Miocene or early Pliocene formation, more likely the former. I found at one spot, evidently washed out from the bank, a large fragment of the jaw of a cetacean. Philippi's Creek is reported to harbor numerous alligators, but on our trip both up and down the stream we saw but a single individual, and that a young animal. The weather was not very warm, and possibly the reptiles may have kept beneath the surface.

A water-way through the mangroves conducts from Big Sarasota Bay to Little Sarasota Bay, and may be used with much advantage by small craft. Owing to the chances of stranding we were compelled to take the outside route, and thus to pass the bars at both inlets. A considerable surf was rolling at the time we entered Little Sarasota Inlet just before sun-down, but we succeeded in making the point, and anchored under the lee of the bar of sand that separates the inlet from the sea, in one of the most picturesque spots that we had thus far seen in our journey.

The rock guarding the entrance to the channel on the north side is a coquina, very similar to that found on Philippi's Creek. It is rapidly undergoing destruction through the wash of the sea, and will, doubtless, in a very short time be completely removed. In color it differs essentially from the typical coquina of the east coast, which is very light, or nearly white, whereas this one is by contrast rather dark.

On White Beach, on the inner side of the bay, we again found large quantities of dead fish strewn over the shore. The same burden rested on the long line of oyster reef which extends not very far from this point into the bay, where thousands upon thousands of carcasses were heaped up in continuous banks, upon which the gorged turkey-buzzards were lazily attempting to recover from their revels. The air was actually foul with the odor of decomposition. A reef rock, of Miocene or early Pliocene age, I was unable to determine which, with numerous impressions or casts of corals, some of them identical with the forms found at Whittaker's, juts out on White Beach, where it has been largely honey-combed through the wash of the water, and in places is rendered soft and friable; in other spots, again, it is tough and very resisting. Among the numerous molluscan remains there were few that were retained in anything like a perfect state of preservation, and scarcely one that permitted of specific determination. Indeed, I only indicate with doubt the occurrence of *Pecten Jeffersonius, P. Madisonius*, and *Venus alveata*. In a somewhat different rock, but without doubt belonging to the same series, we found abundant casts of a large oyster, not unlikely *Ostrea Virginica*, associated with similar remains of the clam (*Venus Mortoni ?*), cockle (*Cardium magnum ?*) and a *Perna*. A small stream empties into the bay near this point, exposing heavy beds of rock on either bank to a thickness of some eight to ten, or twelve feet. I found a few casts of gasteropods in these deposits, and a few fragments of scallops, apparently *Pecten Madisonius*, but the fossils were not numerous, and barely determinable. The difficulty of wading in the stream, too, prevented me from penetrating very far. A short distance from this point we were conducted to a locality where the carapace of a large fossil turtle, measuring nearly three feet across, was embedded in the roadway, of which it formed a part. The time-honored passage of vehicles over it had completely crushed the carapace, breaking in the top, but the outline was still clearly defined in its entire circumference. I secured two large fragments, from which I had hoped to determine the specimen on my return, but, unfortunately, they were left behind at one of our packing stations.

Mr. Brock, who, in company with the cook, had during the absence of the remainder of the party explored a portion of North Creek, another tributary of the bay, reported the existence of a highly fossiliferous stratum exposed on the banks of that stream at an elevation of some ten to twelve feet. This stratum, which is underlaid by a white friable limestone, was traced for a distance of about three-quarters of a mile, but it is said to extend very much further. It is much to be regretted that want of time did not permit us to make a more extended exploration of this very interesting locality, and to definitely determine the different ages of the deposits occurring here. The shell bed is either Pliocene or Post-Pliocene, but the very limited number of fossils that were brought to me for determination, among which I recognized the giant *Fasciolaria gigantea*, pear-conch (*Fulgur perversus*), and clam (*Venus mercenaria?*), did not permit me to settle the question. I strongly incline to the Pliocene age of the deposit, inasmuch as we subsequently found the same fossils occupying a nearly similar position along the upper Caloosahatchie, and in a stratum whose Pliocene age was placed beyond question. Still, from this correspondence alone, I should not like to pronounce too positively on the matter of correlation.

From Little Sarasota Inlet to Casey's Pass the ocean front is made up of a vast shell bank, three to five feet or more in thickness—a non-indurated coquina, if so it might be termed—which at the time of our visit was being rapidly destroyed through the action of the surf. The beach was strewn with dead shells, among which I in vain searched for a living specimen. We dragged in twenty feet of water, but the dredge struck on an unproductive shell-bottom, and brought principally fragments to the surface. The dredge was again thrown over just beyond Casey's Pass, bringing up fragments of arenaceous and serpuloid rock, besides numerous dead shells, principally of the genera Arca, Cardita, and Venus, the greater number of which were stained pink through some peculiar process of ferric oxydation. We also obtained several branches of an Oculina, still retaining much of the colored animal substance or cœnosarc, which would go far toward confirming the assertion of our captain that a submerged coral reef exists opposite this point at a distance of a few miles from the coast. None of the coral-polyps were visible in the mass. We dragged again off Stump's Pass, in water of 10–15 feet, and obtained among other things a beautiful assortment of the large sand star-fish, *Luidia clathrata*.

LITTLE AND BIG GASPARILLA INLETS.—We made Little Gasparilla Inlet on the afternoon of Feb. 24th, anchoring for the night. This is considered to be one of the best collecting grounds on the coast, and our explorations on the following morning fully confirmed this impression, at least so far as our own personal experiences would permit us to form a judgment. The numerous shoals and grass flats, protected and exposed bayous or inlets, afford an almost endless variety of retreats to the different animal forms that abound here, and serve in great measure to circumscribe the individual habitats. Thus, one spot would be largely relegated to a species of Cerithium (*C. muscarum*), another to a second species of the same genus (*C. ferrugineum*), and a third to an association of both these forms. In one of the inlets I found large quantities of the green shells of *Fasciolaria tulipa* inhabited by the *Clibanarius vittatus*, the combined colony, · as if with a common impulse, moving in one given direction. The correspondence existing between the color-tints of the hermit and that of its protecting shell was very remarkable, but whether this correspondence was in the present instance merely accidental or as the result of selection, I am not prepared to say. That a unity of color between the shell and the crab would in a measure tend to conceal the latter from general observation and thus secure for it a partial protection from its enemies, is undeniably true; but it may be questioned whether the peculiar tints of the animal were not, in this special instance, a development depending upon the general surroundings—the grassy bottom, etc.— rather than a relation holding with the shell, the choice and subsequent habitation of which may have been purely fortuitous circumstances.

The Vermetus "reef" was here again largely developed, forming a prominent fringe along the shore margin. I picked up two stranded jelly-fishes, of the genus Cyanea, which had evidently only quite recently been washed on the beach; the disk of the larger individual measured 22 inches in diameter. Both specimens were kept on deck of our schooner for four days, with the object of drying and ultimate preservation; but at the end of that time, owing to an unfortunate accident, which resulted in their partial destruction, and the steadily growing odor of decomposition, I reluctantly heaved them overboard. The elimination of water had been very rapid during the period of desiccation, and in a short time, doubtless, but for the accident, both disks, beautifully exhibiting all the lines of structure, would have been ready for final preservation.

The bottom of the inlet was in places covered with a species of sea-anemone, one of the forms occurring off Sand Key, in Clearwater Bay, and also with the common sea-urchin (*Toxopneustes variegatus*). The latter had in nearly all cases covered itself with a dome of gravel and broken shell—in imitation of the general character of the bottom— which was supported on the extremities of the ambulacral feet, and served to conceal the animal from view. Mr. Willcox had on a previous occasion called attention to this remarkable habit on the part of the urchin, but he seems not to have fully recognized the importance of the deception played by it as a factor in its own defense. So complete was this deception that I must have wandered probably over a full acre of urchin-ground before I was made aware of the presence of these animals; indeed, were it not for accidentally stumbling over one of the hillocks, thereby exposing the animal beneath, I might to the present time have been left in ignorance of their existence there. To positively test the nature of this covering of broken shell I partially filled my collecting bucket with shell fragments, and placed in it a number of the urchins stripped of their covering. With wonderful rapidity the frightened creatures bored their way into the mass of debris, and appeared almost immediately with a large accumulation of shell fragments centred on their ambulacral tips. There could be no doubt, whatever, as to at least one of the uses of this, to some persons purely "ornamental," armor. The shell fragments I found supported indiscriminately by both their convex and concave surfaces.

Mr. Willcox and the cook were very fortunate in securing with the dip-net some half-dozen specimens of a large spotted Aplysia or sea-hare, which appears to be new to science. The largest individual measured about eight inches in length, and full five inches in width. The color of the mantle was sea-green, tinged with purple, with large irregular blotches of lighter color, and numerous white, or at least very light, spots of about the dimensions of the cross-section of a slate-pencil. The nearest ally of this animal appears to be the *Aplysia depilans* (*leporina*) of the Mediterranean, from which, however, the species differs in many essential particulars. I would propose for the new form the name of *Aplysia Willcoxi*. When placed in a bucket of water, especially when irritated, the animal emitted a magnificent purple-crimson fluid, which almost instantly clouded everything in the vessel. Two other species of Aplysia-forms belonging to the genus Notarchus were found at the same locality, one of which appears to be identical with the West Indian *N. Pleii;* the other closely resembles the eastern *N. Savignana*, and may be identical with that species. A dozen or more of the individuals were collected, and placed in our alcohol vessels, the strength of the alcohol in which they were immersed being gradually raised from below 50 per cent. to about 80 per cent. The animals were evidently caught on their feeding-grounds, a grass shoal rising to within about three feet of the water-surface. On our return to this spot, toward the close of our journey, a large individual of the *Aplysia Willcoxi* was observed slowly floating out to sea, propulsion on the surface of the water being effected or assisted through a measured movement of the folds of the mantle.

We found a small sand-fly very abundant at this locality, which annoyed us considerably when on land, the first time that any annoyance was experienced from insect pests. So deficient, indeed, did the entire region thus far traversed appear in insect life that one might almost have concluded that the members of this group were either entirely wanting or but accidentally represented. Travelers who, at this season of the year, expect to meet with a gorgeous entomological display, rivaling what has so frequently been described as a heritage of the tropics, will naturally be disappointed, as will also the botanist, who, in anticipation of the facts of nature, expects to revel in a bed of flowers. It is a mistake to suppose that there are here no true seasons of animal and vegetable life. Hibernation, or retardation of growth as dependent upon seasonal conditions, is probably nearly as well marked in Florida as it is in most of the region situated to the far north, and I have no doubt that the apparent absence of insect life is in reality only a reflection of this period of quiescence.

We dragged off Big Gasparilla and again off Boca Grande, but both times over unproductive grass-bottom.

CHARLOTTE HARBOR.—In the grass bottom off Uzeppa Island, where our schooner anchored for the night, we found numerous single tunicates and a few large clusters of a brilliantly colored branching red-sponge; otherwise there was a marked deficiency in the variety, no less than in the numerical development, of animal life at this place. We dragged opposite the northern extremity of Sanibel Island alternately over a shell and grass-bottom, but the dredge added little of consequence to our collections. An extensive shell-beach faces the ocean front on Sanibel Island opposite to Blind Pass, but at the time of our visit it was strewn almost entirely with dead and water-worn shells; living specimens of the shuttlecock shell (*Pinna muricata*), were, however, very abundant.

We ran aground on a grass shoal just before reaching Punta Rassa, but soon righted, and put into harbor not long after sundown. For hours during this day's journey our vessel was followed by a number of drum-fish, which appear to have hung close to the keel, and whose diabolical serenade was continued from early in the afternoon almost through the night. The different individuals, judged by their "booms," must have retained their relative positions almost without change.

THE CALOOSAHATCHIE.

The region about the Caloosahatchie, and more particularly the interior tract which harbors the headwaters of that stream, are so little known that we found it almost impossible to obtain any information that could prove of advantage to us in our intended exploration of the southern wilderness. The most that could be ascertained was that at certain intervals along the river we would come across settled hamlets or plantations, but the approximate distances at which these furthest outliers of civilization were to be met with were so vaguely stated, and differed so materially among themselves, that it was impossible to place any implicit reliance upon them. No scientific observations, other than those pertaining to pure topography and hydrography, had ever been made in this section of the State before, which fact, coupled with the hope that along this stream we might expect to find a more reliable clue to the true physical history of the State than along any other, provided a geological profile was offered, made us anxious to enter the *terra incognita*. The results obtained amply warranted our determination.

The ascent of the river to Fort Thompson, where a rapid separates the headwaters from the waters of the lower stream, consumed somewhat more than four days, during which time, owing to contrary winds, and the remarkably tortuous channel, frequent recourse had to be had to the pole. The actual distance from the sea-border to the site of this old fort is not more than fifty miles, but measured along the sinuosities of the channel, which are especially well-marked in the upper course, and more particularly in the reach of the last few miles below the rapids, the distance is very nearly twice as great. We found a considerable depth of water, ranging in a general way from about five to fifteen feet, almost along the entire course of the stream, except in the immediate embouchure, or in the stretch of the first few miles above Punta Rassa, where innumerable shoals so completely bar the channel as to render its passage difficult and hazardous to all but the lightest craft. Although drawing but two feet of water, our schooner barely succeeded in effecting an entrance, and on the return journey we were shoaled several times. There seems to be no reason why, with a moderate outlay, this channel could not be so deepened as to permit of a safe and ready entry even for vessels of a moderately high draught, although, manifestly, by reason of the very gradual shelving of the sea-bottom, no really great depth of water could ever be secured. But even with this deepening the tortuousness of the channel would still very materially interfere with the possible conversion of the stream into a highway of travel, and not until connecting canals are cut to shorten distances is it likely that much use will be made of the stream as a water-way either to or from the far interior. The deepest sounding obtained by Engineer Meigs during his official survey of the river was sixteen feet, but at least in one instance, not very far from the site of Fort Denaud, our lead dropped 28 feet, and I am informed by our captain that on a former occasion he had marked off 32 feet. Numerous snags, principally trunks of live-oak and palmetto, around which sand-bars have formed, and are forming, obstruct the channel of the river for a very considerable part of its course, and render navigation in some parts a matter of considerable caution. These could be very readily removed, however, as only in very few places do they appear to be actually jammed.

The width of the stream varies considerably, naturally narrowing very rapidly in its upper course. Here, the numerous projecting or overhanging trees, in their tendency to catch on to the rigging, necessitate a careful rounding of the bights, into which a vessel is apt to be forced by the current of the water. On more than one occasion a pennant, derived from the overhanging vegetation, was added to our topmast, and once we barely escaped serious accident through this novel method of aerial anchorage. Along the lower reaches of the river the mangrove constitutes the predominating element in the vegetation, its dense line of aerial roots forming an impenetrable palisade for miles of the river-front. We found that the plants here had suffered much less from the cold than elsewhere, and they accordingly presented a much more vernal aspect than in the bays and inlets to the north. The foliage was brilliant green, and showed but little of that purple tint which elsewhere recalled our autumnal season. At Fort Myers the orange trees were in both fruit and flower, and here for the first time could we obtain quantities of that most luscious fruit without being compelled to select from a mass of frost-bitten specimens. The general southern limit of the cold wave, which at Tampa is reported to have depressed the thermometer to 18° F., might be said to have been the Caloosahatchie. Still, even along this river many of the more tropical plants appear to have suffered. Thus, while at Fort Myers the cocoanut and date-palm were bearing fruit—noble specimens of their kind—the banana presented a most wilted appearance, the few straggling leaves or stems that were not frost-bitten little recalling those graceful outlines which the delineations of travelers impress upon their sketches of tropical scenery. The pineapple appears to have suffered equally with the banana, both here and further along the river in the interior.

A few miles above Fort Myers the mangrove gradually thins out, and is followed by straggling lines or groups of palmettos, which here attain a height of some thirty to forty feet. Before reaching Telegraph Station, and at intervals beyond, the forest unfolds itself in its full magnificence, the dense tangle of endless creepers and climbers, the rigid but delicate leaves of the palm, whose noble shaft is reared pre-eminent over the forest, and the brilliant greens with which the eye never satiates, forming a picture of scenic loveliness which no pen can adequately describe. The growth along the immediate water margin is very dense, so that in many places no landing can be effected. The almost complete absence of flowering plants was here again very apparent, but I observed at least one species of Ipomæa and a Lobelia in bloom.

One of the largest of the lower clearings is seen at Thorpe's, on the right bank of the river, where, in addition to the cultivation of a number of semi-tropical products, such as the pineapple and banana, there is a considerable industry derived from the growth of the cane, which yields sugar of a fine quality. The soil is reported to be very favorable to the proper development of this vegetable product, which is also cultivated with profit in other sections of the country where but little else is produced. A series of clearings, alternating with larger patches of more or less heavily timbered woodland, ending in a pine tract, extend from Thorpe's to within about twelve or fourteen miles of Ft. Thompson, when an apparently interminable forest of palms clothes the river on both banks. This is probably one of the most extensive tracts of primeval palm growth in the State. The palm trunks range to 40 or 50 feet, or more, in height, and almost by themselves constitute the forest, there being but little intermixture of deciduous trees. There is also little, or no undergrowth, and the eye, accustomed to the impenetrable mazes of the lower river, follows with rapturous delight the beautiful vistas that reach far into the forbidding recesses of the deep interior. Nowhere else did I observe such a wealth of arboreal vegetation; the profusion of plants clustering around the individual palms, forming there aerial gardens of the most fairy-like description, was simply amazing, and, indeed it seemed as though the usual undergrowth of our northern forests had been bodily transported into an upper realm.

The larger game, such as the deer, wolf and American panther, or Florida lion, are said to be fairly abundant in these wilds, especially in the more open country of pines, but we had little opportunity of testing the truth of the currently received notions respecting the distribution of these animals. On the return journey our captain, whom we were compelled to send on a foraging expedition, reported the finding of several deer, but this is the only instance during our entire journey when a mammal, exceeding the raccoon in size, was actually seen, although on one occasion, on the borders of Lake Okeechobee, we heard the cry of a large cat, probably the panther. The birds of the forest were not very numerous either, and they appeared to be restricted to a comparatively small number of distinct types—red-headed woodpecker, cardinal grosbeak, scarlet tanager, a number of warblers, etc. We heard the cackle of the wild turkey on one or two occasions, and once a specimen of this not very rare bird helped to grace our larder. In the open meadow or "prairie" country immediately above Fort Thompson we observed three flocks, of six or eight individuals each, of sand-hill cranes (*Grus pratensis*), whose graceful outlines presented very pleasing objects in relief to the sombre green background. Their utter disregard for our presence and apparent ignorance of any possible injury, even during the firing of a gun, permitted of an easy approach to within short range, but we failed to secure specimens. The only response to our discharge was an aerial saltation of about three feet, followed by a peaceful return to a disturbed, and apparently interminable, meal. From this point inland, the marsh lands, with their scattered "hammaks" of hard-wood, and everglades literally teem with wild-fowl of all descriptions.

We paid but little attention to fishing on the Caloosahatchie, and are therefore not prepared to say much concerning the ichthyic fauna of that river. It is true that we observed, all in all, but a very insignificant number of fishes, but there is reason to believe that the river is fairly well stocked with these animals. The bass and cat-fish are reported to be fished quite extensively, and we caught several specimens of a bream and sun-fish.—The alligator is still fairly abundant in some parts of the stream, especially towards its upper course, but its early destruction is threatened through the endless pursuit of the hide-hunters, whose compensation is about 50 cents for the hide of an animal exceeding five feet in length. The expense of skinning and salting is included in this sum, which, therefore, allows but little margin for profit, and necessitates an appalling destruction of the animal in order to secure the hunter against loss. None of the animals that we saw in the river were of large size, and the greater number probably did not exceed six feet in length.

Owing to the great number of snags in the channel, and the fear of losing our dredge, we were unable to make any systematic observations respecting the invertebrate fauna of the stream; the dark color of the water, moreover, resulting from an infusion of palmetto roots and stocks, limited the range of vision to a very moderate depth, so that we were doubly handicapped. Still, as far as could be ascertained, there appears to be a decided deficiency in this lower fauna. Indeed, almost the only molluscan form that we obtained were a species of Unio, a Neritina (*N. rectivata*), a Planorbis (*P. trivolvis*), and an Ampullaria (*A. depressa*). Other species, doubtless, exist, and possibly even in considerable quantity, concealed along the deeper and inaccessible parts of the stream. In the everglade tract above Fort Thompson two species of Planorbis (*P. trivolvis* and *P.* [*Physa*] *scalaris*), besides the large Ampullaria, were very plentiful, and still nearer the interior lakes the dredge brought up quantities of one or more species of Vivipara (*V. lineata, V. Georgiana?*).

GEOLOGICAL FEATURES OF THE CALOOSAHATCHIE.—The banks of the river for its entire course are very low, at no place probably rising much above twelve feet. They are highest in the middle and upper course of the stream, where their faces are cut down almost vertically to the water's level, below which they descend at a very steep angle. In the lower reaches of the river they barely attain one-half this height, and, indeed, for a very considerable distance above Fort Myers, the average elevation probably does not exceed three or four feet, and beyond the immediate border the land-surface, showing unmistakable signs of periodic overflows, sinks still lower.* Compact rock crops out here and there, or may be seen lying in the channel, but for by far the greater distance the banks consist of a partially indurated marl, in which, at places, fossils are exceedingly abundant. In my experience I have never met with an exposure in which fossils were nearly as plentiful as in the vertical cut which extends almost uninterruptedly for ten or more miles below the Thompson rapids. Fossils could here be counted by the million, and were as densely packed, but without crushing, as it was possible for them to have been placed together. Their state of preservation was also wonderful.

Owing to the innumerable turns in the river, and the fact that the beds exposed maintain a well-defined horizontality for most of their extent, I was unable to satisfy myself as to the direction of true dip†, so that it may yet be an open question how much of the more westerly exposed rocks, or those cropping out at, and immediately above, Fort Myers, correspond to the rocks exposed along the upper stream. The fact, however, that there is such a slight difference in level between the inner and outer points, and the circumstance that for such a long distance the practical horizontality of the beds can be connectedly followed, lead me to suppose that the entire system is in reality one, despite a certain amount of variation both in the lithological and faunal features of the deposits.

* Tide-water, or perhaps more properly back-water, is said to extend to Fort Thompson. We, however, found a strong river-current for a considerable distance below this point, both during our ascent and descent of the river. The difference between mean high water and mean low water at Fort Myers has been determined by Meigs to be 2.2 feet.

† At one point, not very far above Daniels', the strata show a decided declination to the east, or towards the interior of the State, but I feel confident that this marked deviation from the horizontal is a local circumstance, and has but little bearing on the question of true dip.

A tough sand-rock, of undoubtedly recent formation, crops out at Fort Myers, just above the landing; as far as I could determine, it was destitute of organic remains, or when present these were in such a fragmentary condition as to be unrecognizable. I was also shown along the river's bank a number of large nodules or boulders of a fossiliferous limestone, which were reported to have been obtained from a neighboring well-digging. In these the recent *Venus cancellata* was clearly determinable; from the very great abundance of this shell, its excellent state of preservation, and the general appearance of the imbedding matrix, I feel satisfied that the rock is of Post-Pliocene age—certainly not older than late Pliocene. A somewhat similar rock, densely charged with the same species of mollusk, and with various other bivalves, besides a host of gasteropods (Fulgur, Turbinella, etc.) crops out in a field on the left bank of the river, about 20 miles by water above Fort Myers (six or seven miles in a direct line?), not very much beyond Telegraph Creek crossing. The species of mollusk recognized here were: *Venus cancellata*, *Venus mercenaria* (*permagna?*), *Cardita Floridana*, *Arca transversa*, *Fulgur* sp. ?, etc., all of them apparently still living in our waters, from which it is to be inferred that the deposit is of Post-Pliocene age. The rock is overlaid by a sandstone, in appearance identical with that which crops out at Fort Myers, of which it is the probable equivalent. Immediately below the fossiliferous stratum first described a tough rock, largely charged with shell-fragments, and containing numerous impressions of bivalves, mainly of small size, makes its appearance at water-level, below which it extends for probably several feet. The very unsatisfactory condition of the embedded remains, rendering a positive determination of species impossible, precluded also an absolute determination of the horizon. The sharp line of demarkation separating this deposit from that immediately overlying it, coupled with the knowledge that extensive Pliocene deposits are developed in the further course of the stream, leads me to suspect that this basal rock is also Pliocene, or, at any rate, that it represents a geological period distinct from that which is indicated by the *Venus cancellata* bed.

Just below Thorpe's, and in both banks, a white shell marl rises out of the water to a height of about two and a half or three feet. It contains great quantities of a ponderous flat oyster (*Ostrea meridionalis*), distinct from any of the related forms now living, and of two large scallops—*Pecten comparilis*, and a form, *P. solarioides*, resembling it in general outline, but differing in its much greater size, and in several other peculiarities of structure. Both the oyster and the scallops could be detected in the marl-rock some distance beneath the surface of the water, whence several specimens were obtained by means of the mattock. The *Ostrea Virginica* is also very abundant in the sand rock. On top of this fossiliferous white marl, for which I assume a Pliocene age,* there rests a stratum containing innumerable valves of the *Venus cancellata* (Post-Pliocene).

Mr. Thorpe conducted me to an outcrop of compact sand-rock in a palm "hammak," just back of his sugar-mill, which had much the appearance of the rock exposed at our last section on the river. Its absolute stratigraphical relations with the beds exposed immediately on the river front could not be established, but it is certainly very nearly the newest of the series.

The banks increase in height almost immediately after leaving Thorpe's, but for a considerable distance there is a decided dearth of fossil remains. Stray specimens of the oyster or Pecten appear here and there in the beds, but for miles we found practically nothing. Before reaching Daniels' a compact and highly fossiliferous rock forms the upper moiety of the (right) bank, appearing at an elevation of from four to eight feet above the water. Among the large number of molluscan casts occurring here I recognized those of *Venus cancellata* and of species of Fulgur (*F. perversum?*), Turritella, Cardium, Arca, etc., most of them undeterminable specifically. There can be no question, however, that they represent the forms (Pliocene) which occur in such a beautiful state of preservation a short distance further up the stream, and which, by their vast numbers and large size, constitute probably one of the most remarkable exposures of fossils to be seen anywhere. In the lower part of the bank above described we found the large oyster associated with many fragments of the scallops already referred to. We also obtained numerous Rangias from the bed immediately underlying the top-sands.

A fine exposure of yellow and buff limestone, averaging about ten feet in height, is presented above Daniels', the different strata of which it is composed apparently dipping to the east; the bottom bed is a compact shell-rock, containing innumerable shell remains, largely fragmentary. I feel confident that the dip observed here is purely local, a possible result of sagging, and that it does not interfere with the general scheme of horizontality that is presented both above and below this point.

A short distance above this locality begin the highly fossiliferous deposits to which reference has already been made, and which extend practically without intermission to Fort Thompson, a distance along the river of some ten to twelve miles. This is without question the most remarkable fossiliferous deposit that has as yet been discovered in the State, and from a purely paleontological standpoint, perhaps the most significant in the entire United States east of the Mississippi River. The fossils, which are about equally distributed between both banks, crop out

* The same oyster and scallops are contained in the unequivocal Pliocene deposits occurring further up the river, occupying approximately the same relative positions.

in almost countless numbers, and attract attention, apart from their prodigious development, by their great variety, large size, and beautiful state of preservation. The whole bank much resembles a fossil shell-beach, and recalled to my mind the wall of shells extending from Little Sarasota Inlet to Casey's Pass. But that this was not its true character is proved by the perfection in which individual shells had retained their outlines—even the most delicate, such as Pyrula (Ficula), showing little or no surf action—and by the great number of forms (Panopæas, Arcas) which still remained in their normal positions, both valves firmly attached—the same as they originally occupied when living.

The number of recent forms occurring here is very great, so that at first glance I scarcely doubted that the formation was of Post-Pliocene age, a conclusion to which I was further led by the absolute freshness of many of the specimens. Closer inspection, however, revealed a host of forms which had no analogues in the recent fauna, and others, again, which, while closely approximating living species—so much so, indeed, as to leave no doubt as to their inter-relationship—yet differed sufficiently to indicate a long period of time during which the modifications, resulting in the distinctive characters of the recent species, were brought about. This relationship between the old and the new fauna is very remarkable, and perhaps nowhere else does the doctrine of transformism or evolution receive stronger support from invertebrate paleontology than here. The lines of derivation through which some of the modern forms have passed are perhaps best seen in the case of one or two species of Arca, which stand in unmistakable proximity to the recent *A. incongrua* and *A. Floridana*, in a large volute as ancestral type of the comparatively rare *Voluta Junonia*, and in a ponderous stromb, which strongly foreshadows the recent *Strombus accipitrinus*. Other cases of relationship and obvious derivation might here be cited, but these will be specially noticed in the descriptions of species.

It is a singular fact that scarcely any of the distinctively Miocene fossils of the Atlantic coast are found here; such of the Miocene species as do occur are with few exceptions forms that still live along the coast. *Per contra*, the new species are as a rule strikingly distinct, even in their broadest characters, from the members of our hitherto ascribed Tertiary faunas, or from the equivalent faunas of the West Indian Islands. It is difficult to conceive of the radical difference existing between this fauna and that which ought to be most nearly related to it, whether the special comparison be made with the faunas occurring on this side of the Atlantic or the other.

The following enumeration of species exhibits the relation existing between the forms now described for the first time and those that had been previously described, fossil and recent:

*Murex imperialis,
* " brevifrons,
Fusus *Caloosaensis*,
Fasciolaria *scalarina*,
* " gigantea,
* " tulipa,
Melongena *subcoronata*,
Fulgur *rapum*,
* " contrarius,
" excavatus,
* " pyrum,
* " pyriformis,
*Nassa vibex,
Turbinella *regina*,
Vasum *horridum*,
Mazzalina *bulbosa*,
Voluta *Floridana*,
Mitra *lineolata*,
Marginella limatula,
*Oliva literata,
* " reticularis,
Columbella *rusticoides*,
*Cancellaria reticulata,
Pleurotoma limatula?
Conus *Tryoni*,
" mercati?
" catenatus?
Strombus *Leidyi*,
* " pugilis,
Cypræa (*Siphocypræa*) *problematica*,
*Pyrula reticulata,
*Natica canrena,
* " duplicata,
*Crucibulum verrucosum,
Crepidula cymbæformis,
* " fornicata,
Turritella *perattenuata*,
" apicalis,
" cingulata,
" mediosulcata,
" subannulata,
*Cerithium atratum?

Cerithium *ornatissimum*,
*Bulla striata,
*Siliqua bidentata,
Panopæa Menardi,
" *Floridana*,
" *navicula*,
Semele *perlamellosa*,
* " variegatum,
*Rangia cyrenoides,
Venus *rugatina*,
* " cancellata,
" Rileyi,
* " Mortoni,
*Artemis discus,
* " elegans,
*Dione (Calliste) gigantea,
* " maculata,
Cardium *Floridanum*,
* " magnum,
* " isocardia,
Hemicardium *columba*,
*Chama arcinella,
" crassa,
Lucina *disciformis*,
* " edentula,
* " Pennsylvanica,
* " Floridana,
* " tigerina,
Carditamera arata,
Arca *scalarina*,
" crassicosta,
* " lienosa,
" aquila,
" plicatura,
" (*Arcoptera*) *aviculæformis*,
*Pectunculus lineatus,
" aratus,
Spondylus *rotundatus*,
*Plicatula ramosa,
Pecten *solarioides*,
" comparilis,
" Mortoni,
* " nodosus,

*Pecten nucleus,   Ostrea *meridionalis*,
Anomia Ruffini,   *   "   Virginica.

Recent species are preceded by an asterisk; the new species are italicized.

It will thus be seen that the relation of recent to extinct species is as 48 to 41, giving a very much higher percentage for living forms than obtains in any of the divisions of our recognized Miocene deposits, even the "Carolinian," which holds a position nearly equivalent to the so-called Mio-Pliocene of Europe. It becomes manifest that this most extensive Floridian exposure represents the Pliocene age—a circumstance interesting, apart from the general bearing which its presence has upon the geology of the State in particular, from the fact that it gives us the first unequivocal indication of the existence of marine Pliocene deposits in the United States east of the Pacific slope.

I made a careful examination of the banks to ascertain if any dividing lines or horizons, characterized by distinct assemblages of organic remains, existed, but failed to discover any such; the fossils appeared to be packed almost indiscriminately, and in several instances when I thought that a certain localization in some species could be detected, the same forms would appear in other parts of the bank, and completely vitiate all my surmises. Only along the top line was there a true differentiation, the uppermost (marine) bed being densely charged with the valves of *Venus cancellata*, largely to the exclusion of the numerous other forms that so eminently serve to define the bank in general. Nor did I succeed in obtaining any extinct species from this topmost stratum, although no true junction line between it and the stratum immediately underlying could be determined. There is no question in my mind that this upper Venus bed, the same as we found it at other points of the river, is of Post-Pliocene age, continuous sedimentation, however, uniting it with the older Pliocene deposits beneath, and obscuring all well-defined faunal lines of separation.

From the observations that have thus far been made respecting the geology of the State, it will be seen that the Tertiary formations follow one another through the peninsula in regular succession from north to south, beginning with the Oligocene (or late Eocene) and ending with the Pliocene. The Post-Pliocene, doubtless, follows as a continuation of the Pliocene south of the Caloosahatchie, probably for a very considerable distance into the everglade region, and possibly nearly to its end. Our observations failed to bring forward a single fact confirmatory of a coral-reef theory of the formation of the peninsula such as had been advocated by Louis Agassiz and Prof. Le Conte; on the contrary, the existence of the heavy fossiliferous deposits about Tampa, on the Manatee, along the tributaries of Big and Little Sarasota Bays, and more particularly those exposed on the Caloosahatchie, conclusively proves that a coral extension to the southern United States, such as had been theoretically set forth, does not exist in fact. To be sure, remains of coral structures, possibly representing even true reefs, were found at various points, as for example at Ballast Point, Hillsboro Bay, and on White Beach, Little Sarasota Bay, but these limited structures are evidently only of local formation, and indicate a period when a fringe of coral developed where, through unfavorable circumstances, probably induced through a lowering of the temperature, structures of a similar kind are no longer represented. In other words, they indicate nothing more or less than is indicated by remains of a like character found in our more northern Miocene deposits —the masses of Astræa, etc., of North Carolina, the James River, and other localities. Along the Caloosahatchie we found only scattered clumps of coral (Astræa, Colpophyllia, Dichocœnia?), measuring possibly eight or ten inches in greatest extent, and nothing that could be taken to indicate an associated reef.

In conformity with the system of nomenclature which I have elsewhere adopted in the classification of the American Tertiary deposits, I would propose to designate the Pliocene series of the Caloosahatchie as the " Floridian," by this name indicating the region where the formation has its furthest, and, as far as we know, only, development. What its precise equivalent among the trans-Atlantic formations, if any such exist, may be, still remains to be determined. Thus far I have been unable to discover any whose fauna can be strictly, or even approximately, correlated with the present one. Besides shells and corals, and a few hypothetical remains which are perhaps to be referred to the class of annelids, the only other invertebrates found in the banks were several more or less perfect specimens of the large urchin, *Echinanthus rosaceus*. Two of the more remarkable of the molluscan forms occurring here are an ark, differing from all known types of the family, whether recent or fossil, in a peculiar anteriorly projecting spout or rostrum, and a cowry, with a singular channeled apex.

For some distance below the Fort Thompson rapids the topmost of the marine deposits exposed on the river—the Post-Pliocene *Venus cancellata* bed already referred to—is seen to be overlaid by a heavy stratum of limestone, in which the remains of fresh-water organisms, Planorbis, Limnea, etc., are very numerously imbedded. This fresh-water limestone, in many places an absolute shell-rock, compact but largely waterworn, can be traced with few breaks to the rapids (and beyond), where it acquires its maximum development, with a thickness of two or two and a-half feet. It here rises from two to four feet above the surface of the water, everywhere overlying the *Venus cancellata* bed, which in turn here and there exposes the older fossiliferous deposits beneath; these, however, are practically all concealed beneath the water's level.

The fresh-water limestone forms the bed-rock of the beautiful "prairie" or meadow land which opens out immediately above Fort Thompson, and which soon passes off into the region of endless swamps and everglades that continue to the Okeechobee wilderness. There can be little question, it appears to me, that this vast area of scattered ponds and swamps marks the site of an ancient continuous, or nearly continuous, body of fresh water, which covered the region in the form of a vast shallow lake, and whose origin is probably to be traced back to the period when the land gradually emerged from the sea. The general configurations of the country, and the broad extent over which the limestone (or its remains) is spread, leave little doubt in my mind as to a former union of the present scattered waters, whose isolation may have been brought about principally as the result of vegetable growths, or of this in combination with actual desiccation.

The limestone has been traced eastward, as reported by Captain Menge, the officer in charge of the dredging operations connected with the Okeechobee Canal, for a considerable number of miles, disappearing * at a depth of five feet two inches beneath the canal surface, about three miles west of Lake Hikpochee. We, ourselves, traced the extension of the limestone for nearly this distance by means of the scattered shell remains (fossils), which at intervals were dredged up from the bottom of the canal. All the molluscan forms occurring in the limestone are identical with species now living in the river, and consist mainly of *Planorbis (Physa) scalaris*, innumerable shells of which, evidently distributed at a period of recent high-water, are scattered over the open tracts, and in crevices on the trunks of trees. I obtained specimens from tree-trunks at an elevation certainly not less than 10 or 12 feet above water-level, but the high-waterline marked on the palmetto trunks—the traces of a recent overflow—was still much above this, probably fully six or eight feet.

* That is to say, had not been traced further, but there can be no question as to its extension beyond this point.

THE OKEECHOBEE WILDERNESS.

Almost immediately after passing out of the cut which the Caloosahatchie has excavated in the limestone at Fort Thompson our schooner was fouled on one of the banks that obstruct the upper channel, and we were compelled to lie over for upwards of an hour. The current was here particularly swift, and it was only after a most determined effort on the part of our captain, who succeeded in beaming up the ship by wedging one of our dingeys under its bow, that we were able to get off at all. We had suddenly missed the channel proper, but the stranding was the first indication we received of our having gone astray, an experience which we had already lived through on more than one occasion during our Florida campaign. The water was literally alive with coots, whose break through the surface echoed from far and near over the solitudes. Large numbers of ducks were also hidden in the sedge.

Prior to the operations of the Florida land improvement company, whose dredgings have succeeded in opening a navigable channel of a few feet depth of water, this point was practically the head of navigation of the river, which here emerges from a vast expanse of almost impenetrable sedge and saw-grass. Light boats, after being transported over the rapids, could still ascend the stream for a distance of several miles, winding tortuously through the mazes in which the stream is ultimately lost. The newly excavated canal follows for some distance the actual line of the river, being merely an extension of the stream, but after passing through Lake Flirt—at the time of our visit scarcely more than a swamp tract largely overgrown with grass, flag, and various waterplants—almost completely leaves the bed of the old stream, which appears here and there meandering through the wilderness of morass, and pursues a more nearly direct course to Lake Hikpochee, over a total distance of some seventeen miles.

The depth of water in the canal varied from about four to six feet, while the current was running at the rate of probably not less than two miles an hour, if not more. We had the advantage of a favorable wind, and made the passage before nightfall, keeping hard on to the bank over which our boom felled the grass like so much broken chaff. As far as the eye could reach this almost boundless expanse of grass, relieved at intervals by oases of the most luxuriant verdure of palms and cypresses,

Cypress Swamp.

constituted the landscape; the general growth was about six to eight feet in height, dense to impenetrability, but in some places it was very much higher, and completely shut off from view all but the narrowest vistas. We found here a virtual paradise for birds. The red-winged starling, merrily contesting with the more sombre crow-blackbird a peaceful habitation along the immediate banks of the canal, appeared in almost countless numbers, pealing forth a continuous and perpetual strain of song, while hundreds of herons, egrets, and ibises, decked in the majesty of their full plumes, disported among the inner recesses of the morass, or flecked with so many specks of white the clumps of trees that had been selected for their heronries. An occasional limpkin or courlan (*Aramus pictus*) might be observed hovering over a mud-flat, but more commonly its presence is announced through a peculiar distressing cry, from which, not inaptly, the bird has received the name of screamer. On our return journey over the same ground we observed, associated with the white herons and ibises, two flocks of the roseate spoon-bill, a bird not exactly uncommon in these regions, and known to the inhabitants as the "pink curlew."

We were much surprised at the abundance of alligators, whose freshly made, or but recently deserted, "beds" appeared all along the banks. At intervals of almost every few hundred feet one of these grim monsters of mail, disturbed by our approach, would rise, and suddenly turning upon itself, plunge from its sunny retreat into the cooler shades below, disappearing only to reappear after the disturbing element had passed. A limited number of individuals, especially young forms, took no notice of our approach, retaining an air of composure in their siestas which seemingly no ordinary incident could disturb; but the greater number of the individuals took to flight apparently before our approach had actually been noticed, except in so far as it had been announced by the displaced water of the boat spreading commotion in advance of our own coming. As many as six or seven of these animals could at one time be observed from our vessel, lazily crossing and recrossing the canal, sometimes submerged to the extent that only the extremity of the snout and the large eyes were visible, at other times extended out on the surface for nearly their full extent. It is not often, I believe, that one has an opportunity of observing these animals attack their prey, but I was fortunate on one occasion to detect a small alligator seize a turtle by its protruded neck, and draw it beneath the water. This is the only instance during our entire trip that one of these animals was observed in the act of feeding.

The shallowness of the water in the canal permitted us to make considerable use of our landing-net, which, however, brought nothing to the surface but the few freshwater mollusks, recent and fossil, whose species made up the bulk of the limestone exposed at Fort Thompson. The bottom is very largely overgrown with a species of Myriophyllum (?), which is especially abundant on the sands. We dragged in Lake Flirt, but failed to detect anything of significance in the mass of black vegetable muck with which our dredge came loaded to the surface.

The elevation above sea level of the east end of the canal, or where the canal issues from Lake Hikpochee, is 20–22 feet, or about 11 feet above the base of operations near Fort Thompson. This would give a fall of 10–12 feet in a course of some fifteen miles, an average of somewhat less than a foot to the mile. There can be no question, it appears to me, that Lake Hikpochee was the true source, beyond head-springs, of the Caloosahatchie, although, as I am informed by Captain Menge, it was found impossible, during the survey of the canal-route, to trace that river into the lake, the farthest accessible point on the stream, where it eventually loses itself in the maze of saw-grass, being still removed some distance from its western border. Doubtless, however, the water of the saw-grass country is in large part an oozing-out product derived from the lake, just as the waters of the more southern Everglades represent a similar outflow from Lake Okeechobee. Indeed, in view of the very nearly uniform level occupied by the two lakes, and the swamp character of the intervening territory, it is more than probable, despite the existence of a low dividing ridge, that the last named lake is itself, whether directly or indirectly, the most important contributor to the river's basin, largely regulating the height of its waters, and of those of the smaller sister lake lying to the west.

We traversed the lake (Hikpochee) in a direction slightly north of east, at a point where its width was estimated to be about seven miles. The north shore was visible for much of the distance, but in the south no bounding line could be detected. It is remarkable, in view of the broad extent of this beautiful sheet of water, that even as late as 1875 its very existence should have been doubted. The following quotation is taken from Dr. Kenworthy's narrative of a journey in Southern Florida, published in Hallock's "Camp Life in Florida" (p. 298–9): "An examination of Drew's and Colton's maps will show a large lake existing at Fort Thompson, and another some miles east, named Hickpochee. These bodies of water only exist in the imagination of map-makers . . . We instituted many inquiries of Indians, settlers and cattle-drivers regarding Lake Hickpochee, but all scouted the idea of its existence."

We took numerous soundings, which gave an average depth of water along the line of passage of upwards of ten feet, the lead at no place indicating a drop of over fifteen feet. The bottom appeared to be largely overgrown with the same plant which we had observed in the canal, and which, in its profuse development, prevented the dredge from reaching the actual bottom of the lake. I have little doubt, that the true fundament is a compact sand, similar to that which we subsequently found to constitute the floor of Lake Okeechobee, although along the eastern border of the lake, especially at the mouth of the Okeechobee canal, a vast accumulation of black vegetable muck, containing much woody-fibre, and representing the incipient stages of peat formation, clogged the waters over considerable areas. The fauna of the lake appears to be a very deficient one, if we may be allowed to judge from the character of our drags. Apart from a few Unios and Paludinas obtained in one of the western bays, the dredge, in several trials, brought to the surface from deepest water (fifteen feet) only the red larvæ of a species of annelid, a form which was also subsequently obtained in Lake Okeechobee. The fact, however, that the dredge in most instances did not completely penetrate the grass-growth covering the bottom, accounts in a measure for the poverty of the catch; but yet the almost total absence—indeed, it might practically be said, total absence—of animal forms in the grass with which the net came up loaded, is certainly very surprising, and argues strongly for an actual deficiency in the lake fauna. Several species of fish, among them the bass, were fairly abundant in the eastern shallows, where we also obtained a specimen of the alligator-gar, and a number of alligators. None of the last named animals were observed to pass far into the lake.

Contrary winds, and a powerful current, probably not less than three miles an hour in the inflowing canal, prevented us from continuing our journey during the day beyond the eastern margin of the lake. We tried the plan of warping, *i. e.*, pulling the boat by means of a long line doubled over advanced stakes, but were compelled to desist after a drag of a few hundred feet, and after very nearly meeting with a serious mishap. Only four miles intervened between us and the large body of water which so many before us had vainly attempted to reach, and concerning which so many vague and contradictory reports had been spread. I allowed myself to be hoisted to the mast-head, whence, with the aid of a powerful field-glass, I obtained an unbroken survey of the surroundings. To the north and east the eye wandered over an almost unbroken expanse of swamp low-land—here and there a few clumps of hard wood relieving the monotony of the endless sea of saw-grass—while to the west the low line of sedge, making the western boundary of the lake, could just be distinguished. I had expected to obtain a fairly good view of Okeechobee, but a lowering sky, combined with an intervening fringe of willow scrub, practically shut out the object of my search, although from an occasional momentary shimmer I could just determine the position of the ruffled waters of the lake, and mark a boundary to the dreary waste of Everglades.

The next morning, with a favorable wind, we made the connecting passage in less than two hours. The waters of the canal teemed with alligators, and we also observed several turtles sunning on the bank. At the west end of the canal we obtained specimens of *Ampullaria depressa*, *Limnea columella*, *Physa gyrina*, and *Sphærium stamineum*, and also a variety of freshwater shrimp; and at the eastern extremity *Unio Buckleyi*, *U. amygdalum*, and *Paludina lineata*. The last two, in addition to Ampullaria, were also brought up by the dredge from about its middle course.

LAKE OKEECHOBEE.

The exploration of the Okeechobee region consumed the better part of six days, during which time we made a traverse or examination of probably not much less than one-half the area covered by the lake. Our course from the mouth of the canal, which is marked by a fairly conspicuous barrel-shouldered cypress, was S. by E. to Observation Island, about seven miles, two miles S. W. to the western shore, fifteen miles almost due north to beyond the mouth of Fish-Eating Creek, on the northwest shore, fifteen miles E. N. E. to the mouth of Taylor's Creek, which forms the extreme northern (northeastern) apex of the lake, two miles W. to Eagle Bay, and twenty-two miles S. W. to the canal. The distances here given are those of dead reckoning, but the experience of our captain in calculations of this kind leads me to suppose that the figures are not very far removed from the truth. It will thus be seen that our direct examination was confined principally to the western and northern sections of the lake, but from our position at Taylor's Creek we had a clear sweep of some ten additional miles of the eastern shore as well. How much further to the south beyond the furthest point reached by us the lake extends, I am unprepared to say; nor can I determine this question from any of the numerous hypothetically constructed maps of the region. It is, I believe, safe to say that there is not a single map that represents with even approximate correctness the contours of this vast body of water; indeed, the majority of the maps published, and not less, the descriptions, run so wide of the mark in their delineations, that practically no reliance can be placed upon them. And this criticism applies equally to the maps published with the sanction of the State or General Government and those prepared in the interest of special land or railroad companies. Thus, on nearly all the maps the mouth of the drainage canal is represented as opening considerably to the south of the median line of the lake, while Observation Island is located immediately abreast of this opening, or even considerably to the north of it! It has already been seen that the true position of the island is several miles to the southeast of the canal. The limited time at our command, unfortunately, did not permit us to establish the exact position of the canal-opening, but that it could not be much, if anything, below the middle of the lake, is conclusively shown by the open water-way which extends miles beyond Observation Island. Again, on such maps where the position of Observation Island relative to the canal is in a measure correctly located, two other islands, one of which is the Observation Island of most cartographers, figure north of the canal; neither of these islands could we discover, nor do they appear to exist in fact, unless under the name island it is intended to include sundry island-like clumps of willows and cypress which at intervals break off from, or lie under the lee of, the shore. The extreme length of the lake is generally conceived to be upwards of forty miles, and on some maps, *e. g.*, the United States Land Office map of 1882, Granville's map of 1886, is placed as high as fifty miles. Both of these figures I believe to be largely in excess of the truth, although, from our failure to reach the southern extremity, I might be debarred from making a positive statement to that effect. But every indication leads in the direction of exaggeration in the generally received figures. That the delineated dimensions, or the dimensions taken between well ascertained points, are entirely illusory there can be no doubt. Thus, on the two maps above mentioned, the position approximately corresponding with, or intended to represent, the opening of the canal is placed nearly thirty miles south of the northern apex of the lake, Taylor's Creek; whereas, as a matter of fact, the diagonal distance uniting these two points, as measured by our dead-reckoning, was certainly not more than some twenty or twenty-two miles. Making the necessary allowance for this shrinkage in the northern half of the lake, and granting the correctness of the southern half as delineated, the total length would scarcely exceed thirty-six or thirty-seven miles. My own impression, however, is, that the lake is still considerably shorter, probably not very much over thirty miles. As to the greatest width of the lake I can offer no opinion, not having seen the eastern shore except along the northeast border.

Another error, freely perpetuated on our maps, is the location of the mouth of the Kissimmee River, which is made to correspond with the northern or northeastern apex of the lake. This, as has already been seen, is occupied by a broad bayou known as Taylor's Creek, which is distant a considerable number of miles to the east or northeast of the Kissimmee. The closed or obscured opening of the latter stream, which is in a grass country, renders it difficult to find, whereas the boundaries of Taylor's Creek are sharply defined by opposing walls of noble cypresses, which from their great height, 125 feet or more, present the appearance from a distance of low bluffs. The break in the shore-line is here very distinct, and is apparent at a distance of several miles; hence, by some navigators of the lake the opening is mistaken for the mouth of the Kissimmee, and, doubtless, frequently officially reported as such.*

* Mr. F. A. Ober (Fred. Beverly), in his narrative of the "Okeechobee Expedition,"

PHYSICAL FEATURES OF THE LAKE. The lake may perhaps best be described as a vast shallow pan of freshwater, which probably nowhere much exceeds twenty feet in depth. We took numerous soundings all along our course, probably fifty or more, which gave an average depth ranging from about seven to ten feet. The deepest sounding, made on the diagonal connecting Taylor's Creek and the mouth of the canal, about four miles S. W. of Eagle Bay, gave fifteen feet, but this is the only instance where we obtained this depth. Captain Strobhar, however, informs me that on a previous occasion, and not very far from the same spot, he obtained 22 feet. There is good reason to believe, seeing the general uniformity of the bottom, that this figure represents the approximate extreme depth of the lake, and that only at very exceptional intervals does this amount of depression in the basin obtain.

Practically, therefore, the bottom represents a flat plain, elevated some 7–15 feet—in places less—above sea-level. The same plain is manifestly continued into the floor of Lake Hikpochee—which, as has been seen, has the approximate depth of Lake Okeechobee—and, doubtless, forms also the true fundament to the vast series of swamps and everglades which on all sides surround these two larger bodies of water. We sounded at many points in the channels running into the grass and in the cypress thickets, and usually found a considerable depth of water, 6–8 feet, or even more, and where the bottom was reached in these shallows it consisted almost invariably of vegetable muck, of which there appears to be a heavy accumulation, and not of the solid siliceous sand which we everywhere found to constitute the floor of Okeechobee. I think it may be safely assumed that this vast lacustrine plain of the Floridian peninsula represents, practically unmodified, the surface of the country as it appeared at the time of its latest (or only) emergence from the sea. Whether or not a salt-water lake was formed immediately after the elevation of the land, from which through gradual alteration and a steady indraught of fresh-water, the present lacustrine system of waters was ultimately developed, I am unable to say, although the presumption would probably be that there was no such formation. Yet it is not exactly impossible that the reverse was the case. We failed

published in "Camp Life in Florida" (1876), states that "Taylor's Creek, and another smaller, empty into the lake within ten miles of the Kissimmee, but their channels are so choked with water-lettuce and lilies that an experienced eye is required to discern them" (p. 251). What the "smaller" stream may be it is difficult to say; but surely Mr. Ober could not have properly identified Taylor's Creek, when he refers to the difficulty of determining its channel. The high belt of cypress on either side marks it out absolutely. The broad sheet of water at the time of our visit was entirely destitute of lettuce at its mouth, nor does it seem possible that it could ever be seriously clogged at its junction with the lake. Mr. Ober's references to the contours of the lake are exceedingly vague, and in a manner contradictory, so that little dependence can be placed upon them. Fish-Eating Creek is erroneously said to empty into the lake almost opposite Observation Island!

to detect any salinity in the water, which is fairly potable, nor did we discover the remotest traces of any saliferous deposit. On the other hand, however, the valves of *Venus cancellata* were thrown up in considerable numbers both along the beach of Observation Island and near the mouth of Taylor's Creek, and I also succeeded in scooping up, by means of the landing-net, a fairly large fragment of *Fulgur perversus*, and a single shark's vertebra. The shells were all badly worn, and had more the appearance of the specimens contained in the banks of the Caloosahatchie than of the living form, and I am, hence, inclined to the opinion that they represent fossils rather than living specimens. They may have been washed out of the Post-Pliocene *Venus cancellata* bed, which almost positively underlies the lake, buried some distance beneath the sand. There appears to be, however, no means of absolutely determining this point. That the marine animals above mentioned may have succeeded in introducing themselves at a comparatively recent period, after the complete formation of the fresh-water lake, is just barely possible, but very unlikely. In our numerous drags we failed to bring up a single living marine type of animal, nor even a fragment that could reasonably be referred to a living animal of such type—unless, indeed, the numerous individuals of a species of Pandalus, a caridid shrimp common to the waters, be taken to indicate such an organization.* This shrimp was also found in Lake Hikpochee.

It is frequently conceived, and as often reported, that Lake Okeechobee is a vast swampy lagoon, or inundated mud-flat, the miasmatic emanations arising from which render access to it a matter of considerable risk or caution. This is very far from being its true character. The lake proper is a clear expanse of water, apparently entirely free of mud-shallows, and resting, as has already been stated, on a firm bed of sand. All our soundings and drags indicate that this sand is almost wholly destitute of aluminous matter, and nowhere, except on the immediate borders, where there is a considerable outwash of decomposed and decomposing vegetable substances, is there a semblance to a muddy bottom. The water itself, when not disturbed, is fairly clear, and practically agreeable—although held in bad repute by the few who have visited its shores—and by the greater number of our party it was used in preference to the barrel-water with which the schooner was provided. More generally, however, it is tossed into majestic billows, which rake up the bottom, and bring to the surface a considerable infusion of sand, rendering the surface murky. Steadily blowing winds are frequent, presaging heavy swells; we were compelled to lie at anchor for an

* A diminutive shell, much resembling in outline certain forms of Bythinella, but with a longitudinally costulated surface, was sufficiently plentiful in the grass brought up by the dredge; its affinities could not be definitely determined.

entire day during one of these high seas, when the waves beat most unmercifully against our little craft.

The border line of the lake is in most places not absolutely defined, owing to a continuous passage of the open waters into those of the Everglades; on the whole, however, the delimitation of the latter region is fairly well marked, the growth of saw-grass or flag terminating rather abruptly. Where the Everglades constitute the border line, which is the case for the greater part of the west coast, there is necessarily no true shore, and, indeed, it is the common supposition that no landing can be effected in such a region. This supposition is, doubtless, true in its general application, but not absolutely so. We secured a landing opposite Observation Island at a spot where the vegetable accumulation, living and dead, of flag, lily, and grass was so dense as to permit of a safe footing, although numerous holes and pit-falls everywhere revealed the unstable character of the fundament. A pole could readily be thrust into this vegetable bottom to a depth of four or five feet, or even more.

For some distance along the north shore, but more particularly on the northeast, there is a true beach line, made up of oceanic sand. This beach extends for nearly two miles almost due west of the mouth of Taylor's Creek, and probably not less than eight or ten miles, if not considerably further, to the southeast of that stream. It shelves very gradually into the lake, and rises out of it with the same moderate slope. At the localities visited by us I nowhere found it to rise more than about four or five feet above the surface of the water, although not unlikely it may attain a greater elevation. It everywhere supports a dense growth of hard woods —oak, maple, palmetto, etc.—which form a fringe to the almost interminable expanse of saw-grass and cypress-swamp which follows at a very moderate distance in the rear.

OBSERVATION ISLAND.—This island, which lies a few miles S. by E. or S. E. of the mouth of the canal, is perhaps the largest island in the lake, although not impossibly some larger island may exist in the southern bayous not yet explored. It is currently, and even officially, reported to be some two miles in length, but I much doubt if its greatest (north and south) expanse greatly exceeds a half-mile, or, at the utmost, three-quarters of a mile. Along its western and southern borders it is well-nigh inaccessible, owing to a heavy growth of small cypress and custard-apple (?), whose gnarled stems and stumps form an effective barrier to approach. On the east, as also on the north, there is a much more open sand-beach, on which there was a considerable break of water at the time of my visit. The width of the island is at all points very insignificant, and the elevation probably nowhere exceeds four or five feet.

Numerous birds take shelter in the almost inaccessible recesses of this water-bound islet, which is reputed to be one of the most favored of the Floridian heronries. We observed towards night-fall large flocks of the white ibis migrating hither, and similar departures in early morning. The great white heron and the egret were also sufficiently plentiful, but perhaps less so than the water-turkey or snake-bird, whose stoical but uncouth presence gave life to the miniature wilderness.

The only other animals beyond birds collected on the island were a few insects, a scorpion, several centipedes (Iulus), and species of Limnea, Planorbis, Physa, and Cyclas.

TAYLOR'S CREEK.—We spent somewhat more than two days in the exploration of this stream, anchoring a short distance above its mouth in eight feet of water. The width of the channel is at this point several hundred feet, and remains uniform, with a nearly uniform depth of water, for not much less than a half-mile, or even more, beyond which it gradually begins to contract, but without shallowing to any extent. In how much this "creek" is a true creek in the ordinary acceptation of the word, or a simple bayou opening out from the lake, we were unable to determine, owing to the vast masses of floating vegetation, water-lettuce principally, which choke the different channels in their upper courses, and permitted a furthest penetration to our skiffs of probably not more than two or two and a half miles. I found an unmistakable outward current during my first ascent of the creek at a distance of over a mile from its mouth, and up to the furthest point reached by me, but whether this was a natural current, or one produced as the result of direct wind action, or as depending upon a recession of the waters of the lake, could not be satisfactorily ascertained. During my second ascent, on the day following, the water over the same stretch, or over a part of it, was either stationary or slightly receding in the opposite direction. There can be no doubt, whatever, that the direction of flow up to the farthest point reached by us is largely influenced by the condition of the lake—the rise and fall of its waters as depending upon wind action, and not impossibly, also, tidal influences. The absence of shore-lines and of other necessary data rendered impossible, during the short period of our stay, the determination of the actual existence of tidal action in the lake. From a periodic rise and fall of the water in the mouth of Taylor's Creek, measuring some eight or ten inches, but which did not occur at equal periods of time, I am inclined to believe that such action does exist, although the question can, perhaps, still best be considered an open one.

The great body of Taylor's Creek opens out from the lake northward for about three-quarters of a mile, or a full mile, is then deflected northwestward, and after about a quarter of a mile divides into two main arms or branches, one of which is directed to the west and the other considerably more to the north. But no direct course is maintained by either of these branches for any great distance. It is not exactly impossible that other branches, choked at the time of our visit, may open out at seasons into the main channel of equal value with the above, which we were unable to discover. The creek receives three important accessions from the east before the first deflection above indicated.

Nowhere along that portion of the creek explored by us did we find a true bank or shore, the water on either side spilling off into the vast expanse of forest-swamp, principally cypress, which here opens out from the lake. The heaviest timber growth is along the eastern tributaries and immediately about the mouth of the creek, where the parallel walls of majestic cypresses, draped from top to bottom in their funereal hangings of Spanish moss, and towering to a nearly uniform height of 125-150 feet, exhibit to surprising advantage the sylvan wonders of this primeval solitude. It would be vain to attempt to depict by word the solemn grandeur of these untrodden wilds, the dark recesses, almost untouched by the light of day, that peer forbiddingly into a wealth of boundless green—or to convey to the mind a true conception of the exuberance of vegetable life that is here presented. At no time before our visit had I been so thoroughly impressed with the wild grandeur of an untrodden wilderness—nowhere where I so keenly appreciated the insignificance of my own humble being in the sea of life by which I was surrounded.

I made several attempts to penetrate the maze of waters that constitute the "floor" of the forest, and out of which the latter rises, but found the tree-trunks and cypress knees almost everywhere too numerous, rendering it impossible to direct the skiff. The water was uniformly limpid, and nowhere did it appear to be covered with a miasmatic scum of vegetable organisms. Large fields of lettuce float freely on its surface, impelled in given directions by the ever-changing currents that sweep through the interior; where heavily packed these floating gardens are practically impenetrable, and readily carry with them obstacles of a movable character, such as a boat, that might be caught in their path.

The predominating trees of these swamps are the cypress, bay, live-oak, water-oak, and maple, which together impart the physiognomy to the vegetation; occasional palms appear here and there in the less secluded parts of the forest, but evidently neither the watery bottom, nor the exclusion of light which the dense overhanging canopy of interlaced branches affords, is favorable to their development. Although the trees rise to a very considerable height, but few of them attain to really great dimensions. The majority of the larger cypresses do not exceed five or six feet in diameter, and the vast bulk of the trees measure still less; an oak, the largest tree seen on the creek, was estimated to measure about eleven feet a few feet above the roots. A remarkable climbing plant, much recalling in habit the cipó matador of the South American forests, accompanies many of the larger trunks very nearly to their loftiest crown, holding them in a tight embrace, but apparently without exercising much compression, or causing any great discomfort to its host. There are usually one or two coils on a trunk, from which ponderous cables, measuring as much as eight or ten inches in diameter, and tapering inferiorly, depend in straight, or very nearly straight, lines to the bottom. I was unfortunately unable to identify any leaves as belonging to this plant, which possibly ascends as a feeble climber from below, and attains its great expansion in its upper course. The absolutely smooth trunk is grayish-white, and of a still lighter shade than that of the water-oak.*

Animal life is very prolific in these wilds, and at almost all times the forest resounds with the echoes of some of its more musical denizens— the shrill cry of the limpkin or screamer, the hoarse croak of the great blue heron, or the castanet rattle of that amphibious multitude, the frogs, whose orchestration appears never to be final. Towards eventide, when the hooting of the great owl bids the sun to hie, and calls forth the slumbering voices of the night, the dryadic music attains its highest pitch; once more the castanet rattle, and finally all is quiet, save the hoarse bellowing of the alligator, which breaks from far and near upon the stillness of the midnight air.

The larger birds, such as the herons, snake-birds, and ibises are very abundant, but the smaller forms were at the time of our visit conspicuous by their absence. We found no trace of either the roseate spoon-bill or the flamingo, although not impossibly both are found here at certain seasons of the year; the latter is said to breed along the southern borders of the lake. The only time that we met with the spoon-bill was during our traverse of the Okeechobee canal, in the Everglade region between Sugar-berry and Coffee-mill hammocks. We observed several flocks, of some ten to thirty individuals each, of parakeets on Taylor's Creek, and obtained one specimen. These birds frequent the loftiest branches of the forest, calling attention to their gambols by the garrulous tones which they unremittingly send forth.

We met with no quadrupeds in the region, although the tracks of deer and of a large cat, possibly the lynx, were fairly abundant on the sand beach which marks the entrance to the Creek. On one occasion we also heard the distant cry of what appeared to be the puma or Florida lion. Many of the smaller quadrupeds, doubtless, are found here, and possibly even in considerable numbers, but we had no occasion to come across their tracks.

---

* Prof. Gray has kindly directed my attention to the habits of Clusia, to which not unlikely the plant above described belongs. It appears, however, to be very distinct from *Clusia flava*, and may, therefore, represent a species not hitherto described as a member of the American flora.

The bass is sufficiently plentiful in the Creek, and probably constitutes a considerable part of the food of the alligator, which literally swarms here. We were more than astonished at the vast numbers of these creatures, which could be seen or heard at almost every point—here lazily swimming on the surface, there reclining on an intermatted bank, or again splashing unseen from a bed of lettuce and flag. We observed at one time from the deck of our boat no less than nine of these animals unconcernedly swimming in quest of prey, crossing and recrossing the stream in the most methodical manner, suddenly disappearing on an alarm of danger, but reappearing after a brief interval of complete immersion. During my first ascent of the stream, which probably consumed in the neighborhood of five hours, I must have seen or heard in my immediate proximity between fifty and seventy-five alligators, and not improbably many more. They appear especially plentiful at about the middle of day, when the elevated temperature calls them from their aqueous homes. They delight in the masses of floating vegetation that hang matted together on the shore line, whence they can readily see their prey without discovering their own presence. Their power of perception is very acute, and in probably nine cases out of ten, as far as our own experience was concerned, they observed intruders long before they themselves were detected. In no instance did they manifest a disposition to give battle, even when approached to within short range of the boat; on more than one occasion I was sufficiently near to have struck them with a medium-sized pole, or even with the paddle, but the reptiles seemed to entertain no disposition to attack, preferring the easy victory bought by a general immersion. At the same time, they do not always appear to shrink from man's presence, as frequently I observed them heading directly for my boat, disappearing only when so close as to cover me with their downward splash. They are exceedingly tenacious of life, and will execute apparently conscious movements sometimes hours after the head will have been battered in by both ball and axe, the method of execution practiced here. In how far these movements are in the nature of reflex action, excited by some extraneous stimuli, it is not always easy to determine, but in many cases they are without doubt strictly coördinated. On one occasion where I was compelled to use one of our dingeys, containing a young alligator supposed to have been dead for a number of hours, for the purpose of collecting a wounded anhinga, I was surprised, on lifting the bird into the boat, to find the alligator suddenly come to life, and make a dashing onslaught on its unfortunate victim.

The Floridians frequently speak of two varieties of alligator, the red-eyed, which is reported to be the more savage, and the common black-eyed. We observed several individuals of the former, which is also distinguished by a lighter-colored armor, but unfortunately none of the specimens actually obtained by us seemed to show the distinguishing character—or, in other words, all of them were of the common type. Not improbably, as suggested by Dr. Leidy, the red-eye is an albinistic variety. It must be observed, however, that the red-eyed variety in swimming appears in its whole length on the surface of the water, whereas the common form has usually only its nose and eyes, or the head and a portion of the convex body exposed; at least, this was our observation. It is just possible, although not very probable, that we have two distinct species of alligator represented in these southern wilds, and if the crocodile occasionally makes its appearance in Biscayne Bay, why may not also the cayman, or another of its South American congeners? We searched among our specimens for a crocodile, but without success.

The largest alligator killed by us measured about ten feet, but the greater number were much below this figure. We observed, however, several of considerably larger size, and one whose length was estimated at between fourteen and fifteen feet. From several of the individuals we took a number of the peculiar mouth-infesting leech which the species harbors, and from the stomach of one a wholly undigested young bass, measuring about three inches.

The only other reptiles observed in this region were a few individuals of the goitered-lizard (Anolis), and a species of water-snake, apparently new to science, which I picked up in a lettuce-bonnet in Eagle Bay, about two miles west of the mouth of Taylor's Creek.

A few words relative to the ophidian fauna of the peninsula may perhaps not be amiss in this place. It is the common belief that snakes are very abundant in the State, and that traveling in the forest or bush region is made dangerous through their presence. In how far this may be true I am unable to say, but our experience seems to indicate that the abundance of these reptiles, of both the venomous and non-venomous species, is not nearly as great as is currently reported. It is true that our explorations were mainly confined to a border-tract country, and largely to a region of swamp and water, but yet we saw sufficient of the mainland to permit us to form a general notion as to the occurrence of these animals. The total number of snakes seen by us during our entire explorations was about eight, of which at least two were the common black snake (*Bascanion constrictor*), one the water-snake above mentioned, and the remainder in greater part moccasins. Mr. Willcox, who remained along the west coast several weeks after the completion of our general explorations, observed three or four additional individuals along the Homosassa, two of which (moccasins) were secured. We found no trace of the much-dreaded rattlesnake, although the sad experience of a member of a hunting party of the year previous only too truly proves its presence here. The arid sand tracts covered with a dense growth of saw palmetto are the reputed favorite haunts of this animal, and it is here, probably, that the greatest precaution need be had in traveling. Our first moccasin, the one killed on Perico Island, Big Sarasota Bay, was found on a tract of this kind, coiled on the stalk of one of the palmetto leaves. As has already been said, it manifested no disposition to attack, even after being struck with a rake, and it is the common observation here that, unlike the rattlesnake, this equally venomous serpent rarely provokes an encounter, preferring to remain quiet except under immediate provocation, or when impelled in the direction of a food-supply. While gathering fossils in one of the banks of the Caloosahatchie I was for some time in unconscious proximity to one of these animals, whose head, as I am informed by Mr. Willcox, who accompanied me, was less than two feet of my own. Despite our close range, the reptile made no attempt either to escape or to attack, remaining motionless on the overhanging branch from which it was suspended. All things considered, the danger to travelers in Florida from snake bites is inconsiderable, and probably not much more than in many of the proverbially snake-ridden districts of some of our northern States, Pennsylvania or New York, for example. We certainly met with no trace of that swarm of venomous serpents which Bartram reported issuing from almost every stump, nor is it likely that the somewhat unfavorable period of the year during which our journey was undertaken, the hibernating period, will account for the diversity of our success in snake hunting or snake seeing.

Our anchorage in the mouth of Taylor's Creek was almost the only locality where we were seriously annoyed by mosquito pests, although one of our nights in the Caloosahatchie palm forest was passed to the tunes of the little piper. We were, however, in advance of the mosquito season, May—August, when the air is represented to be thick with this social insect. The general dearth of insect life was astonishing, and far from realizing that we were traveling towards the region of its greatest development, it appeared just the reverse. Only on the water surface, or in the lettuce-bonnets, if we except the mosquitos, was there a semblance to anything like profusion. The spiders were here especially plentiful, representing a number of distinct types (Lycosids, Phalangids, etc.), some of them of remarkable beauty. But the nectar-loving insects of the north, the Lepidoptera and Hymenoptera, were practically entirely wanting, a necessary consequence of the almost total absence of flowering plants. This remarkable paucity in the insect life of the region must doubtless be attributed in great part to the early season, and possibly also in a measure to the effects of the recent cold wave of the north.

We found numerous small mollusks, one or more species of Planorbis (*P. lentus*), Limnea (*L. columella*), Physa (*P. gyrina*), and Sphærium (*S.*

stamineum), attached to the under surfaces of the lettuce bonnets, or to their roots, which also supported numbers of diminutive leeches, and two or more forms of crustaceans, one of them a species of Pandalus. The shells were in nearly all cases very thin, and translucent.

FAUNA OF LAKE OKEECHOBEE.—All our observations tend to belief that the fauna of the lake is a very deficient one, and that it is distinctly uniform for the greater part of its extent. We dragged at intervals all along our traverse, with the result of bringing to the surface scarcely more than a half dozen species of animals. Indeed, if we leave out of account the immediate border line of the lake, the entire catch consisted almost exclusively of two species or varieties of Paludina (*P. Georgiana, P. lineata*), and two or three species of Unio (*U. Buckleyi, U. amygdalum*). These mollusks, together with a minute Bythinella-like gasteropod of uncertain relationship, were exceedingly abundant in the lake just off the mouth of the canal, and the dredge came up laden with their shells. Only a comparative few of the shells were without the animals, and in such cases they were largely water-worn, and decalcified. A large proportion of the living Unios had their umbones eroded. Out toward Observation Island the mollusks became much less numerous, but on the north of the lake, between the mouth of the Kissimmee River and Taylor's Creek, they again became plentiful, especially along the beach line of the latter water, where the shells, in company with those of *Venus cancellata*, already mentioned, were thrown up in considerable abundance. They were also fairly plentiful in the vegetable muck of Eagle Bay. The remaining species of Mollusca observed in the lake were the forms to which reference has already been made as occurring on the shore of Observation Island, and on the leaves and roots of the lettuce-bonnets of both Taylor's Creek and Eagle Bay. They are *Limnea columella, Planorbis lentus, Physa gyrina*, and *Sphærium stamineum*. Off Observation Island I scooped up a rock made up essentially of minute Bythinellæ (?), but whether this was of an absolutely recent formation, or a fragment derived from some hidden fossiliferous deposit, I was unable to determine with positiveness.* The species of shell contained in the rock was identical with that dredged up in association with the Unios and Paludinas.

A large proportion of the hauls brought up considerable numbers of a fresh-water shrimp of the genus Pandalus, identical with the species

---

* The affinities of the little gasteropod are doubtful. The surface of the shell is distinctly costulated, and to this extent different from that of any species of the genus with which I am acquainted. Ober mentions a minutely fossiliferous rock occurring toward the southern border of the lake, which is not unlikely structurally identical with the fragment above mentioned.

found in Lake Hikpochee and the connecting canal, which in its general characters approaches very closely some of the more northerly marine species. In Eagle Bay I collected a solitary young cray-fish, the only specimen of this group of the decapod Crustacea observed during our entire trip. The red larva of a species of annelid, measuring about three-quarters of an inch in length, is sufficiently abundant in the bed of the lake, but we were unable to discover the adult which it represents. The same species was also abundant in Lake Hikpochee, while a slightly differing, emerald-green, form was found in the canal connecting the two lakes.

Of the vertebrate life of the lake we found but few traces. The only species of fish obtained by us were the black-bass and cat-fish, both of them of good size. A specimen of the latter, obtained some distance out from the mouth of the Kissimmee River, measured about twenty inches in length. It appears to be specifically distinct from any of the described forms, and I have accordingly proposed for it the name of Okeechobee cat (*Ictalurus Okeechobeensis*). We found the bass very plentiful just at the entrance to Eagle Bay, where the fish were readily caught by means of the trolling line. This method of fishing was also tried for a long time on the open expanse of the lake, but without success. We observed here at intervals a larger fish jump from the water, but the impossibility of a near approach prevented us from ascertaining the species; not improbably it was a sturgeon.

We found the alligator nowhere about the lake, except on its immediate border line—as in the lagoons opposite Observation Island, or in Eagle Bay. This condition was also observed in the case of Lake Hikpochee. Whether these animals perform long journeys by water, or not, I am unable to say, but as far as our own observations go, it would appear that they do not. I noticed two individuals off the mouth of Taylor's Creek swimming leisurely in the lake at a distance of perhaps two or three hundred feet from the actual border.

⁎ The remarkable parasite described on page 46, and doubtfully referred to Clusia, is, I am informed by Mr. A. H. Curtiss, of Jacksonville, Fla., a species of fig (*Ficus aurea*). It is said to at first feed on other trees, "but finally, by sending down multitudes of intergrafting roots, it completely enwraps and smothers the supporting tree and forms a hollow trunk of its own."

RÉSUMÉ OF GEOLOGICAL OBSERVATIONS, AND THE GEOLOGY OF THE STATE.

Prior to our visit the only portion of the State that had been examined geologically, or on which a geological report had been prepared, was the region lying north of a line running almost due northeast from the Manatee River, just south of Tampa Bay, to the east coast. Below this all was conjectural, although the existence of certain limestones of undetermined age was hinted at, or even located, by a number of casual observers (Tuomey, Conrad) who chanced to navigate some of the outer waters. Such a limestone was reported by Tuomey to be found in Charlotte Harbor, but the exact locality of its occurrence is not noted.

The first critical observations on the geology of the State were made by Conrad, who in 1846 described a limited number of fossils from a limestone found in the neighborhood of Tampa, which he correctly assigned to the Tertiary period. This is the earliest positive reference we possess of a definite formation occurring in the State. From the presence of a supposed nummulite in the limestone in question, *Nummulites Floridanus*, now shown to be in reality an orbitolite, Conrad correlated the deposit with the Vicksburg formation of Alabama and Mississippi, from which also a supposed nummulite, *Nummulites (Orbitoides) Mantelli*, had been described, and which had until then passed under the name of the American nummulitic formation. Although the determinations of both of the foraminiferal species have been proved to be erroneous, the correlation of the respective beds containing the fossils remains approximately correct, even if not absolutely so.

It is remarkable that in spite of the positive assurance given by Conrad of the existence of this Tampa limestone so little account should have been made of it in the subsequent theories regarding the formation of the Floridian peninsula, and that the seductive coral theory of growth advanced by Louis Agassiz, and so beautifully formulated by Professor Joseph Le Conte, should have been allowed the weight which it carried for a quarter of a century. Singularly enough, not even Conrad appears to have protested against the theory which had for its ground-work only an examination of the belt of coral reef and the coral keys which border the State on the south and southeast.

The only serious contribution to the geology of Florida made between the date of the publication of Conrad's paper and 1880 is the report of personal observations made in the northern half of the State by Dr. Eugene A. Smith, State Geologist of Alabama, published in the American Journal of Science for 1881. This assiduous investigator collected largely in the way of rocks and fossils, and at localities sufficiently removed from one another as to permit of a broad insight into the geology of the region. From his observations it appeared that at least a considerable part of the northern half of the State, instead of representing a recent formation, as was supposed by many, was in reality underlaid by heavy beds of fairly ancient limestone, which in the characters of their organic remains recalled the upper white limestone of Alabama. The fossils, which were kindly placed in my hands for determination, proved the correctness of the inference as to their age. I identified among them two or more species of Foraminifera, one a large orbitoide, of very nearly the dimensions of the *Orbitoides Mantelli*, representing it, and undeniably the analogue of the European *Orbitoides ephippium*, and an Operculina, very nearly related to, if not identical with, the *O. complanata*. Some of the rock specimens submitted to me were made up of practically nothing but the tests of the aforesaid orbitoide, and of a smaller species more of the type of *O. Fortisii (dispansa)*. From the examination of these specimens I had no hesitation in referring the rocks holding them to the Oligocene formation. The localities noted by Dr. Smith for the occurrence of this formation are located in the tract included between Jackson Co. on the west, and Marion Co. on the south, and it was correctly conjectured that over much, or most, of the intermediate region where no observations had as yet been made, or where there were no outcrops, the same rock would be found as the underlying structure. From one locality, Rock Spring, Orange Co., Dr. Smith obtained rock specimens which differed very essentially in both lithological and faunal features from the specimens obtained at the other localities, indicating the existence of a distinct formation. Among the fossils I identified a number clearly indicative of the Miocene age of the formations—such as *Pecten Madisonius, Cardita arata, C. granulata, Venus alveata*, etc., fixing the most southerly extension of the Medial Tertiary formations of the Atlantic slope. This patch of Miocene is not improbably continuous with the Miocene area of southern Georgia.

The reference of the greater part of the northern half of the State to the Oligocene formation has been more than justified in the light of subsequent investigation, which has disclosed the existence of the older Tertiary rock in many new localities—on the Wacasassa, at Archer, Arredonda, on the Homosassa, Cheeshowiska, etc. From the Cheeshowiska, some four miles from its mouth, Mr. Joseph Willcox, in 1882, obtained large masses of rock, densely charged with foraminiferal remains, among which I recognized great numbers of the Orbitoides occurring in the rock-fragments submitted to me by Dr. Smith, and what was of far greater moment, vast quantities of a true nummulite (named *Nummulites Willcoxi*), the first that had hitherto been discovered on the American continent. The nummulitic masses were embedded in, or bordered by, a fringe of a much newer fresh-water limestone, containing the remains of several recent species of non-marine mollusks—*Vivipara Waltonii, Glandina parallela, Ampullaria depressa*—indicating that there had been a working over of the older formation, and that the specimens obtained were not found *in situ*. Nevertheless, there could no longer be any doubt as to the existence of a true nummulitic formation in the United States, and the age which it represented in the peninsula of Florida. Other specimens obtained by Mr. Willcox at Wacasassa, in Levy Co., contained the remains of two species of sea-urchin, *Euspatangus Clevei* and *E. Antillarum*, identical with forms occurring in the equivalent (Oligocene) deposits of the island of St. Bartholomew.

Since the discovery of these nummulitic rocks on the Cheeshowiska River Mr. Willcox has obtained further specimens of the same foraminifer at a locality removed some fifteen miles northeast of the original locality, and *in situ;* and I have identified the species in rock fragments sent to me for determination by Prof. A. G. Wetherby, from a well-boring situated five miles S. W. of Gainesville. A second species of the genus (*Nummulites Floridensis*) has also been described by me from Hernando Co., associated in a rock mass with various other species of Foraminifera, of the genera Heterostegina, Biloculina, Triloculina, etc.

Concerning the (marine) Tertiary deposits newer than the Oligocene, the only positive indication that we had of their existence in the State prior to 1886, beyond the patch of Miocene above referred to as occurring at Rock Spring, was furnished by Dall (Science, VI, p. 82), who, in July, 1885, reported the discovery of *Ecphora quadricostata*, a characteristic Miocene fossil of the Atlantic border of the United States, in the rock of Tampa Point. Mr. L. C. Johnson about this time also indicated the occurrence of what appeared to be newer Tertiary deposits in the northwest of the peninsula, a conclusion at which I had likewise arrived from an examination of a limited collection of fossils obtained from Ballast Point, on Hillsboro Bay, a few miles south of Tampa.*

During the spring of 1884 the late Prof. W. C. Kerr, of the North Carolina Geological Survey, made a cursory examination of the region about Tampa, the results of which are embodied in a brief paper published

---

* Mr. Johnson (Science, May, 1885) does not give the paleontological evidence on which the rock of Hawthorne and Waldo is referred to the Miocene (or later) period; and I fail to see the full force of the argument which places it newer than the Oligocene, although this may be so.

in the Journal of the Elisha Mitchell Scientific Society for the years 1884–85 (Raleigh, N. C.), pp. 86–90. In this paper the author describes the limestones of Tampa Bay, and sundry outcrops of rock on Six Mile Creek, Alefia River (Bloomingdale), Manatee River (Rocky Bluff), and Peace Creek (Fort Meade), which are all referred to the Upper Eocene period. The general conclusions are thus stated: "The limestone rock underlying the region of country about Tampa belongs to the upper Eocene, as already pointed out by Conrad and Tuomey. And as a gentleman of intelligence who visited Fort Myers, informed the writer that the rock at that place was both in appearance and in fossils, similar to that about Tampa, the Eocene limestone rock almost certainly extends at least as far south as that point." It will thus be seen that the conclusions reached by this observer are absolutely at variance with the facts which our investigations have brought out; and it is not a little surprising that Prof. Kerr should have failed to recognize the post-Eocene age of the Rocky Bluff limestone, when the Miocene character of the fossils occurring at that locality is so clearly marked.

One of the most interesting contributions to the history of the extinct fauna of the State that had been made up to this time was the discovery by Dr. J. C. Neal, near Archer, Alachua Co., of an extensive series of mammalian remains, referable probably to several distinct geological periods, from among which Dr. Leidy has identified the parts belonging to one or more species of mastodon, rhinoceros, tapir, horse (*Hippotherium ingenuum*), llama, camel, and hog. One of the forms of mastodon appears to be distinct from any of the other species inhabiting the United States, and has been named *Mastodon Floridanus*. The hog is considered to represent the type of a distinct genus, *Eusyodon*, considerably removed from the peccary-forms, Dicotyles, Platygonus, which hitherto alone represented the Suidæ in the Tertiary and Post-Tertiary deposits of the North American continent.

The above sketch represents in brief the condition of our geological knowledge of the State up to 1886. It recognizes the existence of Tertiary (Oligocene and Miocene) deposits in the upper half of the State, or north of a line connecting Tampa Bay on the west with Sanford, on the Atlantic slope, on the east, but leaves, as has already been intimated, everything conjectural south of that line. From the broad extent of the Oligocene rocks in the north Dr. Smith was led to infer that in all probability the greater part of the peninsula, at least as far as the Caloosahatchie and Lake Okeechobee, was underlaid by the same formation, the hypothetical extension of which is indicated on the map accompanying his report in the American Journal of Science above mentioned. The conclusion reached here was a natural one, and is that which guided me in the delineation of the geological boundaries on the general map of the Tertiary deposits accompanying my "Contributions to the Tertiary Geology and Paleontology of the United States" (1884). Our recent explorations, however, prove the utter erroneousness of the view here entertained, and indicate not only that the Oligocene formation is in great part, if not wholly, absent from the southern half of the peninsula, where it is replaced by the Miocene and Pliocene, but that even in the north its proper limits, as far as the determination of an absolute horizon is concerned, have probably not been satisfactorily ascertained. Thus, the foraminiferal rock of Tampa, containing the supposed nummulite, $N.$ (*Orbitolites*) *Floridanus*, and through which it had been correlated with the Vicksburg formation, is very different from the true orbitoitic or nummulitic rock which is distinctively characteristic of the Oligocene districts of the further north and the interior. Indeed, it contains nothing in common with this rock, but very much that is distinctive of itself, and the underlying siliceous rock that forms the floor of Hillsboro River. It may, nevertheless, be Oligocene, but in that case it in all probability represents a higher horizon than the same formation further to the north. The reasons for considering the formation as of Miocene age, and as the partial equivalent of the medial Tertiary deposits of Santo Domingo, are fully set forth in pages 121-22 of this report.

A few words only need here be said with reference to the theory, advocated by Louis Agassiz and Prof. Joseph Le Conte, which held that the peninsula of Florida was of comparatively recent formation, and that it represented in greater part merely an accumulation of successive or consecutive coral reefs. Our observations, which had already been preceded in the northern part of the peninsula by the researches of Eugene A. Smith, clearly demonstrate the erroneousness of the views hypothetically set forth, and establish beyond a question of doubt that the progressive growth of the peninsula, as far, at least, as Lake Okeechokee, and probably considerably further, was the result of successive accessions of organic and inorganic material, brought into place through the normal methods of sedimentation and upheaval. A full exposition of Prof. Agassiz's views is embodied in his "Report on the Florida Reefs," published in the Memoirs of the Museum of Comparative Zoology for 1880, some twenty-nine years after its preparation. In this paper the author strenuously denies all evidences of upheaval over the greater part of the peninsula, and ascribes the varying elevations, at least of the more southerly portions, to upthrows or accumulations of material as depending upon the agency of gales and high-water. Prof. Le Conte, while still adhering to the fundamental ideas expressed in his original paper, published in the American Journal of Science for 1857, in a more recent paper (Science, Dec. 14, 1883) disclaims the agency of coral growths in the formation of the peninsula north of the Everglades.

ROCKS OF HOMOSASSA RIVER.—A good section of the rock of the region is exposed at Wheeler's, on the left bank of the stream, about one mile above its mouth. It is a tough limestone, rising at different stages of the water probably two or three feet above its surface—at the time of our visit about one and a-half or two feet—and exhibiting numerous holes, fissures and sinks that have been irregularly worn into it by the water. The surface appears to be practically horizontal, exhibiting no measurable dip. The rock is densely charged with foraminiferal remains, all, as far as I have been able to determine, referable to the Miliolidæ, or to the group of the imperforate Foraminifera, as distinguished from the Foraminifera Perforata, which, in the rock-masses further to the south, usurp their place. I propose to designate this formation, representing one of the four distinct types of foraminiferal rock found in the State, the "Miliolite Limestone"—a member, doubtless, of the Upper Eocene or Oligocene series, more likely the latter. Unfortunately, the absence of distinctive molluscan remains in the rock rendered the determination of its absolute age a little uncertain, but its close proximity to the unequivocal members of the Oligocene rocks of the Cheeshowiska River, a few miles to the south, leaves little room for doubt as to their near equivalence.

The genera of Foraminifera recognized as occurring in the Miliolite Limestone are:

Biloculina,   Quinqueloculina,
Triloculina,   Sphæroidina.

Another, and less compact, limestone, not unlikely belonging to the same series as the last, is found in the immediate neighborhood of Wheeler's. I examined large blocks of the rock that had been taken from a well-digging, but the numerous fossil impressions contained therein, mainly of bivalves, were so obscure or ill-defined as to permit of no satisfactory results being arrived at from their partial determination.

ROCKS OF THE CHEESHOWISKA RIVER.—These are exposed nearest to the sea on John's Island, at the entrance to the river, where a yellowish, spongy limestone, containing numerous molluscan remains and a few tests of the foraminiferal genus Orbitoides, appears on the ocean front at low-water. The main-rock, which is of Oligocene age, is bordered by, or incased in, a rock (limestone) of newer formation (Post-Pliocene), in which the remains of marine organisms are freely intermingled with those of a fresh-water type, such as Vivipara, Ampullaria, etc., represented by species living at the present day. The re-formation of the old limestone is thus made evident, but there can be no question, seeing the character and structure of the country generally, that the unaltered or mother rock is the rock of the region—*i. e.*, that the Orbitoitic sees its true development here.

The same admixture of rocks is again observed at an outcrop on the right bank of the river, about one mile further up the stream. About three miles above this second point, in a clearing known as Loenecker's, is the famous nummulite locality where Mr. Willcox obtained the original nummulites described by me in 1882. The rock lies here in loose masses in a partially ploughed field, about four to six feet above the surface of the water. No indication of a true outcrop could anywhere be detected, but I am satisfied that the parent rock either immediately underlies the locality, or is found in its immediate vicinity. The very insignificant elevation of the entire region above water-level, its uniform horizontality, and the dense capping of vegetable matter beneath which it lies buried, necessarily reduce to a mininum the chances of finding an outcrop, and explain why to so many of its inhabitants this portion of the State is supposed to be wholly destitute of solid rock. Even along the river-courses, where we should most naturally expect to meet with an exposure, the vegetable growth is so dense and impenetrable as to practically completely hide the fundament, and were it not for an occasional or accidental clearing, such as Loenecker's, one might be left in absolute ignorance of its very existence.

The rock occurring at Loenecker's is of two kinds—one, in which almost the entire mass is made up of the tests of orbitoides, and the other in which the nummulites predominate to about the same extent. But several of the larger fragments indicate the most unmistakable inter-association of the individuals of both these genera of Foraminifera, and leave no doubt as to the equivalence in age of the two classes of deposit. The Orbitoides found here represents two or more species, one of the type of the well-known Biarritz fossil, *O. ephippium* (*O. sella*), whose peculiarly infolded tests are very abundant, and another more nearly recalling *O. dispansa*. I am not sure that I recognized any undoubted *O. Mantelli*, the form of the Mississippi and Alabama "Vicksburg" beds, but not impossibly some of the smaller tests referred to the *O. dispansa* type may represent arrested forms of this species. Both species of nummulites occur here, *i. e., Nummulites Willcoxi* and *N. Floridensis*, the former very largely predominating.

With reference to the physical history of these deposits, I can perhaps not do better than quote my own words published at the time of their discovery (Contributions to the Tertiary Geology and Paleontology of the United States, p. 81, 1884): "As to the age of the formation represented by these nummulitic deposits, there might appear to be at first sight no question of doubt. The presence alone of nummulites in any formation is almost positive indication as to the Eocene or Oligocene age of that formation, and the more especially when the remains of these organisms occur in any abundance. Admitting the supposition of this age, we should naturally look to the associated fossils for further confirmatory evidence bearing on this point. Singularly enough, in the case of the Florida nummulitic rocks—at least in the fragments that have been placed at my disposal—with very few exceptions all the molluscan remains belong to a period much more recent than the Eocene, and to species that are still living at the present day. And what may appear still more singular, they are referable in principal part to land and fresh-water genera, Glandina, Paludina, Ampullaria.* From this association and the circumstance that nummulites are still met with in existing seas, it might readily be inferred that there has been here a commingling of contemporaneous marine and fresh-water organisms, and that the deposits in question were laid down under such conditions—proximity to the mouth of a river—where a commingling of this kind could take place. Indeed, it would be difficult from paleontological evidence alone to disprove such an assumption, were it not that almost incontrovertible proof to the contrary, in addition to that furnished by the *abundance* of nummulites, is afforded in the presence of the remains of Orbitoides, a genus which attained its greatest development in the Upper Eocene (" Nummulitic ") and Oligocene periods, and which does not appear to have survived the Miocene. There can, therefore, be little or no doubt that the rock fragments marked by this admixture of an older and a newer (Post-Pliocene or recent) fauna, and comprising both marine and fresh-water types of organisms, have derived their faunal character in great part from the deposits of a more ancient formation, which formation represents, and is the equivalent of, a portion of the European " Nummulitic " (whether Eocene or Oligocene). The exact locality (or localities) which these Florida nummulitic deposits occupy *in situ* has not yet been ascertained, but it is fair to assume that the beds lie along the Gulf border (possibly in great part submerged), where, through the disintegrating action of the oceanic surf, their fragments have at a comparatively recent period been washed together with the material that at the same time was being carried out by the fresh-water streams. The precise position which the formation holds in the nummulitic scale, as fixed by Hantken or La Harpe (*Études sur les Nummulites du Comté de Nice*, Bull. de la Soc. Vaud. des Sc. Nat., vol. xvi, pp. 223-24, 1879), cannot be positively determined from our present data, since exceptionally the group of the *Nummulites plicatæ* is represented as well in the oldest as in the newest of the Tertiary deposits marked by the members of this class of organisms."

* The species identified with recent forms are *Glandina parallela, Paludina (Vivipara) Wallonii*, and *Ampullaria depressa*.

The above remarks remain applicable to the facts of the present day, and require little or no modification except in so far as more recently obtained data permit us to emphasize with greater positiveness certain points that had hitherto remained in a measure conjectural. Thus, the very extensive inter-association of the nummulites with *Orbitoides ephippium* leaves practically no doubt that the formation in question is not Upper Eocene, but Oligocene; and secondly, the finding of nummulites *in situ* at a locality some fifteen miles northeast of the original locality, and, again, five miles southwest of Gainesville, enables us to locate within definite limits the partial boundaries of the formation.

We were informed that an outcrop of the rock occurs in a morass about a half-mile or more inland from Loenecker's, but the lateness of the hour at the time of our visit, and the difficulty of reaching an unknown spot practically inaccessible in the heart of the wilderness, prevented us from making a search in that direction. About a mile and a-half above Loenecker's a mass of rock juts out from the bottom of the river-channel to within a few inches of the water's surface, in one or two spots rising slightly above it. Much to our surprise we found it to be almost entirely destitute of fossil remains, showing not a trace of either of the genera of Foraminifera so abundant in the rock below. Its stratigraphical relations could not be definitely ascertained, but without much question it is a member of the nummulitic series of deposits, and may be a near equivalent of a similar looking rock that appears on the beach at Clearwater, immediately north of the wharf. The fossil impressions are very obscure, and such as we observed can only doubtfully be referred to Cytherea and Modiola.

ROCKS OF THE PITHLACHASCOOTIE.—I was unable to make a personal exploration of the rocks of this region, and am, therefore, compelled to confine my remarks to an examination of rock specimens brought to me by Mr. Willcox, and to this gentleman's references bearing on their occurrence. The rock is a tough, partially siliceous, white limestone, in places densely charged with fossils. These are mainly in the form of casts of minute gasteropods, probably one or more species of Cerithium, among which I failed to find any species that we had observed elsewhere; the question of age is thus left undetermined, although a strong probability argues in favor of the Oligocene. According to Mr. Willcox the shores of the Pithlachascootie are in places rocky for a distance of two or three miles from the mouth, the bluffs rising 6–8 feet above the water, consequently higher than in the case of most of the western streams. Where the bank rises higher than three or four feet, it slopes back and is covered with soil. The fossiliferous beds were found to extend to a height of about three feet above the surface of the water, appearing in both banks. Mr. Willcox also obtained fragments of a silicco-calcareous rock, containing numerous casts and impressions of one or more forms of sea-urchin, from a small island situated about three miles' southeast of the mouth of the Homosassa River. One of the species (the most abundant form) is, I believe, without question the *Pygorhynchus Gouldii* of Bouvé (Proc. Boston Soc. Nat. Hist., Dec., 1846; *Ibid.*, Jan., 1851. —*Nucleolites Mortoni* of Conrad, Journ. Acad. Nat. Sciences Phila., II, p. 40), from the older Tertiaries (Oligocene?) of Georgia. The same species was subsequently identified by McCrady in the limestone of Alligator (Columbia Co.?), Florida, and by him referred to a new genus *Ravenelia* (Proc. Elliott Soc. Nat. Hist., Charleston, March, 1858).

ROCKS OF THE TAMPA BAY AND HILLSBORO REGIONS.—Two distinct kinds of rock appear at Ballast Point, on Hillsboro Bay, at about water line, rising in most places not over two feet above it, still oftener less, and at Newman's perhaps three or four feet. The one rock, a highly fossiliferous yellow limestone, is manifestly in place, and forms the bed of the beach, shelving at a moderate angle beneath the waters of the bay. It contains numerous impressions of the *Venus penita* described by Conrad in 1846, and large numbers of the singular foraminifer referred by this paleontologist to Nummulites (*N.* [*Nemophora*] *Floridanus*). It is very remarkable, in view of the abundance and perfection in which the fossil occurs at this locality, that an imperfect or abnormal specimen, misleading in the details of its structure, should have served as a type for a description and illustration of the species.

The rock containing this supposed nummulite has generally been referred to the Vicksburg group, but as far as paleontological evidence goes, I see no valid reason for considering it to belong to this age. None of the distinctive Oligocene fossils of the formation occur in the rock, nor did we find in it any traces of the foraminiferal types so characteristic of the Oligocene region of the northern part of the peninsula. On the contrary, all the fossil forms occurring here appear to be distinct, except in so far as they are represented in the second kind of rock above referred to, and in the rock corresponding to it which forms the bed of Hillsboro River. If Oligocene, the rock in all probability represents a horizon higher than the Vicksburg beds and the Florida "Nummulitic" (or Orbitoitic), but the evidence is all but conclusive for considering it Miocene, and as the near equivalent of some of the beds of the island of Santo Domingo. I failed to detect in the rock any traces of distinctively Miocene fossils; but the association near by of an indisputable Miocene deposit leaves little room for doubt as to the absolute relationship.

The second form of rock found at Ballast Point is of much firmer consistency than the limestone, and appears in large, rounded or angular bluish masses scattered over the latter, seemingly representing a newer formation. It contains numerous fossil impressions, partially silicified, a number of them, as the *Venus penita* above mentioned, identical with forms found in the limestone, but the greater number distinct. Among these the remains of one or more species of Cerithium are especially abundant, and might be said to determine the faunal character of the rock. We found no traces of the orbitolite. At no place could I detect a solid outcrop of the rock, and, therefore, from the observations made at this point alone it would be impossible to determine the stratigraphical relations of the two series of deposits occurring here. But along the Hillsboro River and on a small tributary, known as Magbey's Spring, which enters a short distance above the town of Tampa, the relationship is very clearly shown. The hard siliceous blue rock, charged with the remains of Cerithium, etc., appears at scattered intervals all along the river bank, now on one shore then on the other, and manifestly forming the bed of the channel. Just below the shipyard the nearly horizontal strata lie in beds of from one to two feet thickness. I was unable to determine any true dip. The orbitolite limestone is exposed a short piece above this point on Magbey's Spring, about three or four hundred feet from the borders of the Hillsboro, in a heavy mass some seven to ten feet in thickness. Although the irregularity of the outcrop and its small extent prevented me from locating its absolute position, there can be no question, seeing its proximity to, and elevation above the river, that it overlies the blue rock of the channel. This must then also be the* relation existing between the two kinds of deposits exhibited at Ballast Point, which are manifestly the equivalents of the Hillsboro series. The big irregular masses or boulders which here extend into the bay, or lie scattered over the limestone, are evidently exposed as the result of outwash, and appear to have been scattered to their present positions through the action of a heavy sea.

The relations of the coralliferous deposit exposed at Newman's landing, as well as of the two other classes of rock just described, are fully set forth on pp. 120–123, and require little further consideration. An enumeration of the species of fossils occurring here is given on pp. 119–120, and 124. The fossils are nearly all completely silicified, and exhibit to the minutest detail the ornamentation characteristic of the different species. The coral geodes, some of them measuring as much as eight or ten inches across, are especially beautiful, and exhibit to good advantage the mammillated character of the substituting blue and blood chalcedony. There is little doubt in my mind that the formation is due to an infiltration of silica in a heated condition, but in what precise manner the peculiar method of hollowing was brought about I am unable even to guess at. Many of the species so closely resemble recent forms that it is at first sight difficult to distinguish between them, but close comparison in almost all instances reveals some constant characters by means of which the two series can be separated. The number of clearly-marked extinct species is, however, very great, and sufficient to fix approximately, in default of direct stratigraphical evidence, the position in the geological scale which the deposits occupy. This is, without doubt, in the Miocene series, but just along what horizon it is difficult, or even impossible, to determine.

ROCKS OF THE MANATEE RIVER.—The deposits exposed on the right bank of the Manatee River at Rocky Bluff, a few miles above Braidentown, have been referred to in the narrative (p. 13) as consisting of a basal marly limestone, and yellowish sandstone, and an overlying siliceous conglomerate, almost totally devoid of fossil remains. The white marl, on the other hand, is distinctly a shell rock, in which casts of fossils, mainly bivalves, and their impressions, are exceedingly numerous. Among these I recognized several forms distinctive of the Miocene formation of the north, such as *Pecten Madisonius, P. Jeffersonius, Perna maxillata, Venus alveata, Arca idonea* (?), etc., which left no doubt in my mind as to the age of the deposit containing them. Fossils were much less abundant in the accompanying yellow sandrock, but the species represented were practically identical with those of the marl. The latter disappeared after a comparatively short distance, but the sandrock continued in irregular honeycombed ledges to the furthest point reached by us on the river. The total elevation of the exposure is not more than three or four feet above the river's surface.

The discovery of a Miocene formation in this portion of the State was not a little of a surprise, as it completely invalidated all the conjectural ideas that had been framed relative to the geological structure of the peninsula. It confirmed my impression as to the intermediate or equivalent age of the beds occurring near Tampa, and clearly indicated what would in all probability prove to be the true succession of the beds further to the south. In other words, it was made manifest that this portion of the State was neither that recent creation which the upholders of the coral theory of growth had claimed for it, nor of that antiquity which was assumed for it in virtue of the hypothetical extension of the Oligocene beds. On the contrary, the evidence was conclusive that the same physical forces which effected the formation of the newer Tertiary series of the eastern border of the United States were similarly operative on the Gulf coast, and that the peninsula of Florida participated in the same general movements that were known to have affected the United States between Georgia and New Jersey during approximately equivalent periods of time. In how far these movements were of both elevation and subsidence still remains to be determined, but the facts point strongly to the conclusion that the growth of the peninsula southward was a nearly continuous one, without much interrupted sedimentation, or any great break in the chain of organic evolution to mark the successive accessions of territory which the peninsula received during its development.

ROCKS OF SARASOTA BAY.—The marine deposits bordering the sea on Big Sarasota Bay are mainly in the form of indurated sands, or where there has been a sufficient infiltration of iron, of partially compact sandstones. Fossil remains are almost wholly wanting, being limited, as far as our own observations went, almost exclusively to the casts of one or more species of single coral of undetermined relationships. These we found in a semi-compact yellow rock, of about three or four feet thickness, at a locality known as Whittaker's. The rock has the appearance of being a comparatively recent formation, and I should probably unhesitatingly have referred it to the modern epoch were it not for the coral impressions which it contains. For the present I feel some hesitation in assigning to it a definite position, although fairly assured that it is late Tertiary, or, possibly, even Post-Tertiary. The same impressions occur in a much more compact and highly fossiliferous rock of White Beach, Little Sarasota Bay, which in the character of its organic remains seems to occupy a position intermediate between the Miocene and Pliocene series.

The ferruginous sandrock exposed at Hanson's, whence I extracted a part of the skeletal remains (converted into limonite) of man, as well as the more compact terrestrial rock that appears some three-quarters of a mile lower on the bay, have been discussed in the narrative (p. 15), and require no further consideration at this place. The only other localities about the bay where we observed fossiliferous deposits were on Philippi's Creek, an eastern tributary, where a yellow arenaceous limestone, highly charged with fossils, most of them in the form of casts or impressions, and but barely determinable, could be observed at intervals along the shore, rising about two feet out of the water. A number of the fossils appeared to be identical with forms occurring in the yellow rock of the Manatee River; especially was this the case with the corals and polyzoans, but the only species that I could definitely locate were *Pecten Madisonius, P. Jeffersonius*, and possibly also *Arca idonea*. The formation is evidently either Miocene or Pliocene, or one holding a position intermediate between the two. At one or two spots near the mouth of the creek, well observed on the right bank, this rock is seen to be overlaid by a heavy bed of coquina, some three or four feet in thickness, the shell fragments composing which are largely triturated, and only differ from the typical coquina of the east coast in their greater compactness. This is, I believe, the first instance that a recent rock of this kind has been noted as occurring on the west coast. A similar rock, now rapidly undergoing destruction through the wash of the sea, guards the entrance to Little Sarasota Inlet. On White Beach, on the inner side of the inlet, probably two and a-half or three miles from its mouth, a reef-rock, very tough in places, and extensively honeycombed through the action of the water, forms the shore line, and doubtless, also constitutes in greater part the bottom of the bay. Unfortunately, the fossils, which are markedly abundant, are in the main in a very bad state of preservation, and for the most part do not admit of specific determination. The impressions of reef-corals, probably a species of madrepore, are numerous, and we also found several casts of apparently the same species of simple coral which has been noted as occurring in the sandrock at Whittaker's. Most of the molluscan remains are in the form of casts and impressions, and belong chiefly, at least as far as the more prominent forms are concerned, to Pecten, Cardium, Arca, Venus and Turritella, the last being by far the most abundant of the gasteropod genera, and in our own collections almost the only one represented. The rock is either of Miocene or Pliocene age, but I could not positively determine which, although from its position, and in the light of our present knowledge regarding the formations on the Caloosahatchie, I should consider it not far from the junction line of the two series, if it does not, indeed, effect a passage between the two. The only forms that appeared to be recognizable, and even these were somewhat doubtful, were fragments bearing a close resemblance to *Pecten Jeffersonius*, *Venus alveata* and one of the northern forms of Turritella.

In a rock manifestly belonging to the same series, although of a somewhat different lithological aspect, we found numerous casts of one or more species of large oyster, one of them, with little doubt, the *Ostrea Virginiana*, in association with which were the casts also of a cockle (*Cardium magnum?*), clam (*Venus Mortoni?*), and a Perna. Many of these were lying loose on the beach, having been evidently washed out from the parent rock. A short distance beyond this point, where a not exactly insignificant creek has cut a nearly vertical channel, the rock is exposed in heavy beds of from one to two feet thickness, rising to a total height above the creek of some eight or ten feet, or possibly more. Among the fossils gathered here, which were neither numerous nor well-defined, there were a number of gasteropod casts, probably Turritellas, and fragments of a large scallop which bore a strong resemblance to *Pecten Madisonius*.

ROCKS OF THE CALOOSAHATCHIE.—The remarkable series of deposits exposed on this river, which I have designated the "Floridian," and which give us the first unequivocal evidences of the existence of a marine Pliocene formation in the United States east of the Pacific slope, have been fully detailed in the narrative (pp. 27–31), and only require incidental mention in this place. They appear in most places as a partially indurated marl or earthy limestone, of a yellowish, buff, or white color, and either largely destitute of organic remains, or so densely charged with them as to constitute a pure shell-rock. At their first visible outcrop, about twenty miles by water above Fort Myers, they barely reach water level, but they gradually rise higher and higher, until some twenty or twenty-five miles below Fort Thompson, their elevation reaches (or reached at the time of our visit) fully six to eight feet, and this elevation is maintained throughout a considerable part of the nearly continuous exposure of some twelve or fourteen miles that immediately precedes the Fort Thompson rapids. At this locality they, in company with the overlying Post-Pliocene Venus bed, disappear beneath the heavy capping of Fort Thompson fresh-water limestone, fully described in the narrative, but there can be no question that their inward extension is still very much greater.

Inasmuch as the deposits in question have been traced to a point removed by fully 40–50 miles in a direct line from the sea, or to a position one-third across the State, they afford the most conclusive evidence, if any such were still needed, of the utter fallaciousness of the theory that seeks to explain the formation of the peninsula on the assumption of successive coral growths. They, moreover, clearly indicate that one of the last chapters in the history of the formation of the State was practically identical with the series of closing chapters that rounded off the physical history of the eastern border of the United States generally—steady sedimentation, slow and gradual upheaval, and absence of specially disturbing forces which might otherwise have interfered with the regular processes attending local organic evolution.

TAYLOR'S CREEK.

GENERAL SUMMARY AND CONCLUSIONS.

1. The whole State of Florida belongs exclusively to the Tertiary and Post-Tertiary periods of geological time, and consequently, as a defined geographical area, represents the youngest portion of the United States.

2. There is not a particle of evidence sustaining the coral theory of growth of the peninsula; on the contrary, all the facts point conclusively against such theory, and indicate that the progressive growth of the peninsula, at least as far as Lake Okeechobee, has been brought about through successive accessions of organic and inorganic material in the normal (or usual) methods of sedimentation and upheaval. The evidence, further, is very strong that beyond Lake Okeechobee and the Caloosahatchie the structure of the State is for the most part identical with that above it, and the observed facts clearly prove that this correspondence must exist over at least a considerable portion of the unexplored region of the Everglades.

3. The Florida coral tract is evidently limited to a border region of the south and southeast. Fossil corals occur sparingly in the Pliocene and older Tertiary deposits, but their appearance indicates only sporadic cases of coral growth, such as are observed at the present day on the borders of the reef-seas, or marginal and included reefs of limited extent, similar to those found on the Miocene border of the Atlantic States generally (Maryland, Virginia, North and South Carolina).

4. The formations represented in the State are the Oligocene, Miocene, Pliocene and Post-Pliocene, which follow one another in regular succession, beginning with the oldest, from the north to the south, thus clearly indicating the direction of growth of the peninsula. The successive Tertiary belts do not follow a direct east and west course, but appear to be deflected from the west northeastwards, so as to conform more nearly with the Atlantic coast line on the eastern border of the United States. The amount of overlap possibly resulting from deposition on opposed borders could not be ascertained.

5. No indisputable Eocene rocks have thus far been identified in the State, but not improbably some such exist in the more northerly sections, and possibly include even a part of what has generally been referred to the Oligocene. In how far the older formations were overlaid by deposits of a newer date, or to what extent the northern half of the State may have participated in a general submergence coincidently with the formation of the more southerly portions, remains to be determined.

6. Sedimentation and deposition along this portion of the American coast appear to have been practically unbroken or continuous, as is indicated by the gradational union of the different formations, and the absence of broad or distinct lines of faunal separation.

7. The strata as far as could be ascertained are very nearly horizontal, or dip at only a very moderate angle, but no true or direct line of declination could anywhere be detected. At no locality could any two formations be unequivocally identified as resting one above the other, except in so far as the Post-Pliocene represents one of the factors under consideration.

8. No disturbance of any moment, or one sufficient to sensibly react upon the rock masses, seems to have visited the Floridian region since the initial formation of the present State in the Older Tertiary period. The elevation of the peninsula, especially in its more southern parts, appears to have been effected very gradually, judging from the perfect preservation of most of the later fossils, and the normal positions—*i. e.*, the positions which they occupied when living—which many of the species still maintain.

9. The northern half of the State represents in great part a deep-sea formation, whereas the southern half is marked by deposits indicative of a comparatively shallow sea. It would appear that before its final elevation a large part of the peninsula, especially its southern half, was for a considerable period in the condition of a submerged flat or plain, the shallows covering which were most favorably situated for the development of a profuse animal life, and permitted of the accumulation of reef-structures and of vast oyster and scallop banks. The present submerged plain or plateau to the west of the peninsula may be taken to represent this condition. Fresh-water streams, and consequently dry land, existed in the more southern parts of the peninsula during the Pliocene period, as is proved by the interassociation of marine and fluviatile mollusks in the deposits of the Caloosahatchie.

10. The modern fauna of the coast is indisputably a derivative, through successive evolutionary changes, of the pre-existing faunas of the Pliocene and Miocene periods of the same region, and the immediate ancestors of many of the living forms, but slightly differing in specific characters, can be determined among the Pliocene fossils of the Caloosahatchie. The doctrine of evolution thus receives positive, and, most striking, confirmation from the past invertebrate fauna of the Floridian region.

11. Man's great antiquity on the peninsula is established beyond a doubt, and not improbably the fossilized remains found on Sarasota Bay, now wholly converted into limonite, represent the most ancient belongings of man that have ever been discovered.

FOSSILS OF THE PLIOCENE ("FLORIDIAN") FORMATION OF THE CALOOSAHATCHIE.

Murex imperialis, Swainson.
Zoological Illustrations, 2d ser., ii, p. 67.
Tryon, Manual of Conchology, ii, p. 101, pl. 23, fig. 206.

A limited number of individuals from the banks of the Caloosahatchie below Fort Thompson, showing no essential variation from the living form. The occurrence of this species in the Pliocene deposits of the State practically determines its true home to be the Atlantic border of America and not the Pacific coast, as is frequently asserted (Reeve, Conchologia Iconica, Murex, sp. 42).

Murex brevifrons, Lam., *var. calcitrapa,* Lam.
Animaux sans Vertèbres, lx, p. 573.
Tryon, Manual of Conchology, ii, p. 95, pl. 19, fig. 175.

Banks of the Caloosahatchie below Fort Thompson.

All the specimens obtained by us appear to be immature forms, and none would measure, if perfect, more than an inch and a half in length. I can detect no character by which to distinguish them from the recent form, unless it be in a greater regularity and prominence of the revolving lines, and a further projection of the variceal spines. The species also grades into *M. crocatus*, which appears to be nothing more than a nodulose variety of *M. brevifrons*.

Fusus Caloosaensis, nov. sp. Fig. 1.

Shell typically fusiform, of about ten volutions; spire acuminate, about one-third the length of shell, with a slightly papillate apex; whorls sub-angular, longitudinally ribbed, and crossed by somewhat distantly-placed, elevated, revolving lines, the median two of which on the whorls above the body-whorl appear more developed than the others.

Body-whorl with about nine revolving lines, the two on the shoulder, with occasional intermediate finer lines, least prominent; aperture semi-oval, with a long, nearly straight canal, of about twice its own length, the two combined considerably more than one-half the length of the entire shell; outer lip thin, striated internally.

Length, 2.2 inches; width, .6 inch.

Caloosahatchie, in the banks below Fort Thompson.

This shell most nearly resembles among living forms *Fusus Dupetit-Thouarsi*, but differs from it in its subangular whorls (evenly convex in the latter), and the absence of the numerous lines on the shoulders. In the first of these characters, as well as in the smaller number of revolving lines, it also differs from the closely resembling *Fusus Henekeni* of Sowerby, a common Miocene fossil of the West Indies.

Fasciolaria scalarina, nov. sp. Fig. 2.

Shell sub-fusiform, of about ten volutions, longitudinally ribbed or plicated; spire elevated, nearly one-third the length of shell; whorls convex, those of the spire angulated at about the middle, crossed by numerous elevated, and more or less rounded, revolving lines, from ten to fourteen on each of the more prominent whorls; central line most prominent, forming the median angulation or carination; interstitial finer lines present at irregular intervals.

Body-whorl with about 15-20 ribs or plicæ, which, as a rule are less prominently angulated than those of the spire, making the shell appear more regularly convex; revolving lines somewhat over thirty, a limited number of which are interstitial; aperture about one-half the length of shell, or somewhat over, produced anteriorly into a broad canal of moderate length; outer lip prominently lined on the inside; columella with two very oblique folds, the lower of which is practically obsolete.

Length, 6.5; width, 2.5 inches.

In the banks of the Caloosahatchie, below Fort Thompson.

The only American Fasciolaria to which this species bears any great resemblance is *F. Sparrowi* of Emmons (North Carolina Geol. Surv., 1858, p. 253, fig. 115), from the Miocene of North Carolina, or at any rate, a fossil from that State which has been identified as such by Conrad. I find in the collections of the Academy of Natural Sciences of Philadelphia two specimens of a large Fasciolaria marked by Conrad "*F. Sparrowi*, Emmons, N. C.," and have every reason to believe that the forms so identified are Emmons' species, although differing widely from the description given by that geologist. This description is, however, very vague, and manifestly erroneous in several of its details, so that little satisfaction can be derived from it; Emmons' figure more nearly resembles the fossil in question, and this fact, combined with the knowledge that Conrad had access to the collections of the Carolina survey, lead me to assume the correctness of the latter paleontologist's identification.

From the *Fasciolaria Sparrowi*, recognized as such, the Florida fossil differs in the lesser convexity, and more pronounced median angulation of the whorls of the spire, the greater number and prominence of the longitudinal ribs, which are largely obsolete on the body-whorl of *F. Sparrowi*, the greater relative elevation of the spire, and the absence of the regular alternation of coarse and fine revolving lines seen in the Carolina species.

Among recent forms the species apparently most nearly related to *Fasciolaria scalarina* is *F. filamentosa* of Chemnitz, from the Indo-Chinese seas. This species has much the habit of the American form, but is largely deficient in the number of longitudinal ribs, which are also much more distinctly nodose, and bring about a very prominent angulation to the whorls of the spire not less than to the body-whorl; the columellar plaits are three in number, instead of two. The same characters approximately serve to distinguish *F. scalarina* from the European Miocene *F. Tarbelliana* of Grateloup (Atlas Conch. Foss. Bassin de l'Adour, pl. 23, fig. 14; Hörnes, Die fossil. Mollusken d. Tertiär-Beckens von Wien, I, p. 298, pl. 33, figs. 1–4), which, if not identical with the recent form above referred to, is certainly very closely related to it.

Fasciolaria gigantea, Kiener.

Icon. Coq. Viv. p. 5, pl. 10, 11.
Tryon, Manual of Conchology, iii, p. 75, pl. 60, fig. 14–16.

Two specimens of this large conch were found in the upper part of the banks below Fort Thompson, on the verge of the Post-Pliocene layer. They do not seem to differ essentially from the living form. The nodes appear somewhat less prominent, and are in a measure indented through a passing shallow sulcus.

Fasciolaria tulipa, L.

Syst. Nat., 12th ed., p. 1213.
Tryon, Manual of Conchology, iii, p. 74, pl. 59, figs. 1–5.

In the banks of the Caloosahatchie, below Fort Thompson. The largest specimen measures just six inches in length. A probable variety of this species, with a more elevated spire and a correspondingly depressed aperture, appears to be identical with the variety figured by Tuomey and Holmes in their work on the Pliocene fossils of South Carolina, pl. 30, fig. 8.

Melongena subcoronata, nov. sp. Fig. 3.

Shell broadly-turbinate, of about five volutions; spire moderately elevated, scalariform, its rounded whorls profoundly ribbed, subangulated or carinated medially, and crossed by numerous well-defined revolving lines, which alternate as coarser and finer striæ; the ribs of the whorl next to the body-whorl, and sometimes also the one above, distinctly tuberculated or spinose, or even coronated.

Body-whorl very ventricose, obtusely angulated above, with a row (in presumably adult individuals) of from eight to ten prominent scaly tubercles, which stand outward from the shoulder angulation; a single row of supra-basal tubercles, six or seven in number, some of which are developed into short pyramidal spines; entire surface covered with coarse, closely arranged striæ, which may become partially obsolete on the shoulder.

Columellar surface broad, slightly flattened, completely covered by the thin labium; aperture about two-thirds the length of shell, or less, quadrangular, broadly-open.

Length of largest specimen, somewhat more than four inches; greatest width, over the shoulder angulation, 2.7 inches.

Caloosahatchie, in the banks below Fort Thompson.

This species closely resembles the recent *Melongena corona* of the southern coast, and, at first sight, can be readily mistaken for it. But it differs in the greater elevation of the spire, the obtuseness of the shoulder angulation (shoulder concave in *M. corona*), the much smaller number of spines, both on the superior and inferior carinations, the nature of the spines or tubercles, which are much more nearly closed (less scaly) in *M. subcoronata*, and in the circumstance that the shoulder spines are directed outward, and not upward and inward, as we find them in *M. corona*.

Fulgur rapum, nov. sp. Fig. 4.

Shell pyruliform, closely inwound, with a short depressed spire; whorls of the spire about five, gently crenated basally (or above the sutural line); apex papillate.

Body-whorl ventricose, high, convex, sub-angulated above, and to an extent also inferiorly, somewhat nodulose on the rounded shoulders; neither true tubercles nor spines; tendency to nodulation in some cases entirely wanting; aperture of nearly the entire length of the shell, elliptical above, and produced into a long, narrow, straight canal, which is slightly deflected to one side; outer lip strongly lined internally.

Columella arcuate, rapidly contracting the aperture at the beginning of the canal; columellar fold not very prominent. The entire surface of the shell covered with closely-placed, moderately elevated, revolving striæ, which have a gently sinuous outline, and exhibit a distinct alternation of coarser and finer lines.

Length, 6.5 inches; width, 3.5 inches.

In the banks of the Caloosahatchie, below Fort Thompson.

None of the recent species of Fulgur at all approach this shell, and even among the fossil species there is none from which it cannot almost immediately be distinguished. The species that most nearly resembles it is undoubtedly the form figured by Conrad on pl. 47 of his Medial Tertiary Fossils (1839) as *F. maximus*, which has apparently never been described. The ornamentation of this shell is practically identical with that of *F. rapum*, but it can be readily distinguished by its broad and arcuate canal, and the more elevated and scalariform spire. The swelling along the base of the body-whorl in *F. maximus* also helps to identify the species.

Fulgur maximus.
Conrad, Fossils of the Medial Tertiary Formations, pl. 47, not described.
Gill, "On the Genus Fulgur and its Allies," Am. Journ. Conch., iii, p. 46 (enumeration of species only).

Shell sub-pyruliform, with a short scalariform spire; spiral whorls about five, convex, slightly hollowed above the middle, the upper two or three gently carinated and crenulated.

Body-whorl ventricose, somewhat concave on the shoulder, which supports a number of irregularly placed, and not clearly defined nodules; longitudinal lines of growth well-marked, disfiguring the surface of the shell; an irregular swelling near the base of the whorl; aperture nearly four-fifths the length of the shell, oval above, produced into a broad and open arcuate canal; outer lip striated internally.

Columella sigmoidal, its surface covered with a thin callus; columellar fold nearly obsolete. The entire surface of the shell covered with numerous slightly wavy revolving lines, which in a measure alternate in size.

This species, in its typical form, cannot readily be confounded with any of its immediate congeners; the absence of well-defined tubercles serves to distinguish it almost at a glance. But the incipient nodulation seen in some, or most of the specimens, becomes much more sharply defined in others, and, indeed, advances with such gradational steps that a continuous passage is led up to the prominently tuberculated *F. Tritonis* (Conrad, Proc. Acad. Nat. Sciences, 1862, p. 583), from the Miocene of Virginia, and from this again, by insignificant changes, to *F. filosus* (Proc. Acad. Nat. Sciences, p. 286), from the same series of deposits. The gradation is absolute, and permits of no natural separation of the different forms here indicated. Whether or not, therefore, these forms are to be regarded as distinct species, or as varieties representing but a single species, with well-marked characters defining the extreme forms, is of little moment. That they are modifications of, or derivatives from, one and the same form, there is, it appears to me, very little doubt.

This species I identified among the fossils of the Caloosahatchie by a limited number of specimens. The largest of these, which measures upwards of five and a half inches in length, is absolutely undistinguishable in character from the Miocene fossil.

Fulgur contrarius, Conrad.
Am. Journ. Science, xxxix, p. 387; Fossils Medial Tert. Form. U. S., pl. 45, fig. 11.
Busycon perversum, Emmons, North Carolina Geol. Rept. p. 249, fig. 107.

Common in the banks of the Caloosahatchie, below Fort Thompson.

This shell has the general character of *Fulgur rapum*, from which it differs in being sinistral. Dr. Gill, in his review of the genus Fulgur (Am. Journ. Conch., iii, p. 144), remarks that in the greater number of sinistral shells the form is not more obliquely wound than in the dextral, as may be proved by the use of a mirror. This may or may not be true as a general thing, but the rule certainly does not hold in the case of the mutually representative species here referred to. The obliquity in *F. contrarius* is decidedly more pronounced than in *F. rapum*.

I am not exactly satisfied as to the relation which this species holds to the recent forms. It certainly most nearly approaches *F. perversus* of Linnæus, and I must admit that it is undistinguishable from some of the non-spinose varieties that are usually referred to this species. Whether these last, however, are specifically distinct from the typical spined *F. perversus* I am not prepared to say; they certainly have much the same general facies, and they would appear to grade into one another. Still, the distinguishing characters—rounded body-whorl, absence of spines, and a more closely-enveloping spire in the non-typical form—are well-marked, especially in the case of the fossils, where they appear to be constant, and may serve to characterize a good species. But whether distinct or not, it is positive that the fossil is represented in the living fauna. Our collection contains one very ponderous form, which retains all the distinctive characters of the smaller individuals, and may be immediately separated from *F. perversus* by its great convexity, the rounded outline of the body-whorl, and its pear-form.

Dr. Gill (*op. cit.*) erroneously refers to this species the form described by Tuomey and Holmes, in their work on the Pliocene Fossils of South Carolina, as *Busycon perversum* (pl. 29, fig. 3). The species in question is Conrad's *F. adversarius*, referred to (Am. Journ. Conch., iii, p. 185) as the "only reversed form with tubercles instead of spines," but which, as far as I am aware, has thus far never been described.

Fulgur excavatus, Conr.
    Am. Journ. Science, xxxix, p. 387; Foss. Med. Tert. Form., pl. 45, fig. 12.
    Cassidulus Carolinensis, Tuomey and Holmes, Pliocene Foss. S. Carolina, p. 147, pl. 30, fig. 1.

Common on the banks of the Caloosahatchie, below Fort Thompson.

The typical forms of this species can be readily distinguished from the recent *F. pyrum*, Dillw., by its scalariform spine, more depressed, and slender body, and the much deeper sub-sutural canaliculation, which is also carried further towards the apex. The shell, in addition, appears to be considerably thicker. But these characters do not appear to be invariable, and I am far from satisfied that the species ought not to be classed rather with the recent form than as a distinct type. Numerous intermediate stages unmistakably unite it with *F. pyrum*. As is the case with the last, this species also exhibits a marked variation in the amount of angulation of the body-whorl, sometimes appearing merely rounded, with only a faint trace of carination, at other times very sharply angulated, and with a decidedly concave shoulder.

Fulgur pyrum, Dillwyn.
Catalogue, 485.
Tryon, Manual of Conchology, iii, p. 143, figs. 402, 403.

One of the specimens that might be said to constitute the *Fulgur excavatus* series is unquestionably the recent form. It proves the impracticability of drawing closely delimited division lines, where specific characters so closely approach one another, and demonstrates the necessarily arbitrary classification which the evidences of transformism must carry with them.

Fulgur pyriformis, Conr.

I have been unable to find a description of this species, and only know it from a specimen in the Academy's collection marked such in Conrad's handwriting. It is identical with the scalariform varieties (so considered) of *F. pyrum* (*F. plagosus* of Conrad?), which in several characters depart widely from the typical forms of that species, and might, perhaps, with propriety be considered distinct. It stands intermediately between *F. excavatus* and *pyrum*.

Banks of the Caloosahatchie below Fort Thompson.

Turbinella regina, nov. sp. Fig. 5.

Shell ovate-oblong, sub-fusiform; spire elevated, gradually tapering, and consisting of from eight to ten volutions; whorls nearly flat, or slightly convex, somewhat angulated above, and only nodulose in the region of the apex; surface covered with revolving raised lines, about five on each whorl below the upper angulation, above which they are less pronounced and more closely placed.

Body-whorl convex, considerably longer than one-half the length of the shell, and ornamented by numerous raised lines, similar to those found on the other whorls. Toward the base these lines become more crowded, somewhat flexuous and coarse, appearing in the form of paired rugations; suture impressed; aperture elliptical, produced into a straight, but deflected, canal of considerable length.

Columellar surface covered with a thick deposit of callus, which leaves partially uncovered a long and narrow umbilicus; columellar plaits three, the median one of which is the strongest.

Length of longest specimen—imperfect below and above, and lacking probably an inch and a half—eleven inches; width across the centre, four inches.

Caloosahatchie, in the banks below Fort Thompson.

We found but two specimens of this stately Turbinella, which in linear measure surpasses all other species of the genus, with the exception of *T. scolymus*. In its general characters it most approximates among recent forms *Turbinella ovoidea* of Kiener (Icon. Coq. Viv., 7), which is said to inhabit the coast of Bahia, Brazil, and is not an uncommon species in the Miocene deposits of Santo Domingo (Gabb, "Santo Domingo," Trans. Am. Phil. Soc., xv, p. 218), but differs in its much more ponderous proportions, the greater relative elevation of the spire, the absence of well-marked nodulations on the whorls of the spire, and the smaller number of prominent revolving lines on the whorls of the spire. The body-whorl also lacks the basal quadrangulation seen in *T. ossidea*. I know of no fossil form that in any way approaches the Florida species, nor any species of the existing Gulf fauna that is as near to it as *T. ossidea*.

Vasum horridum, nov. sp. Fig. 6.

Shell ovate, thick, ventricose, with the greatest width at about the middle; spire elevated, about one-fourth the length of the shell, and consisting of ? nodulose volutions.

Body-whorl strongly angulated on the shoulder—the angulation being at an angle of about 45 degrees to the outer wall—and probably prominently coronated with foliaceous or lamellar tubercles; surface, as well as that of the rest of the shell, profoundly grooved, with about eight sharply elevated revolving ridges below the shoulder angulation, the sixth and seventh from the top most prominent, and separated from each other by a space equal to two of the other interspaces; sutural line somewhat impressed, and bordered inferiorly by a lamellar ridge; aperture produced posteriorly into a short canal, somewhat more than one-half the length of the shell.

Columellar surface covered with a thick deposit of callus, which leaves partially exposed a broad umbilicus; columellar plaits three, of which only the upper two are prominent, the basal one being rudimentary.

Length of imperfect specimen, lacking probably a full half-inch, about five and a half inches; width, three and a half inches.

Caloosahatchie, in the banks below Fort Thompson.

We obtained but a single adult specimen of this large *Vasum*, which apparently exceeds all other species of the genus in size. Unfortunately, its imperfect condition prevents the absolute determination of all its characters, but sufficient remains to indicate that it is a clearly-defined species. There are but few traces of the coronating tubercles left, but I think there can be no doubt, seeing the acute angulation of the shoulder, and the character of the spines in the young shell, that the tubercles were squamose, as in the East Asiatic *V. imperialis* and not of the typical blunt type seen in our southern *V. muricatum*. From the latter, apart from the character just indicated, the Florida fossil differs in its less turbinate form, the greater length of its spire, the smaller number and much greater prominence of its revolving ridges, and in the deficiency of its columellar plaits. The shoulders of the whorls are also much less depressed than in the recent species, rising at a considerable angle instead of being nearly flat.

In the sum of its characters the Florida fossil appears to be most closely related to *Vasum imperialis*, but unfortunately this assumption is based solely upon an examination of the figures and description of that form, the collections of the Academy of Natural Sciences, otherwise so rich in the department of conchology, being deficient in the species.

Genus MAZZALINA, Conrad.

This genus was constituted by Conrad for the reception of an Eocene fossil from Claiborne, Ala., which in general characters approximates the recent forms now generally referred to Lagena of Schumacher. The genus is thus briefly characterized (Journ. Acad. Nat. Sciences of Philadelphia, 2d ser., iv, p. 295): "Turbinate, smooth; columella projecting interiorly and furnished with closely arranged, oblique, obtuse plaits." Mr. Tryon, in monographing the species of recent shells, appears to have overlooked this description, for in the appendix to his review of the family Fusidæ (Manual of Conchology, iii, p. 225, 1881) he makes the following statement: "Genus Mazzalina, Conrad, not characterized;" —and further: "The type [*M. pyrula*] appears to be very similar to Lagena, Schum., if not identical with that genus. I figure it from the original specimen."

The reference of the genus to Lagena is, I believe, erroneous, and is probably founded upon an imperfect examination of the unique specimen, which is decidedly Lageniform, and the assumption that other specimens, if found, would depart somewhat from the type, and more nearly approximate the recent form. In the light of additional specimens obtained during our recent explorations, I can affirm that the characters, such as they are, separating Mazzalina from Lagena are distinct and permanent, and leave no doubt as to the propriety of separating the two genera (or sub-genera). These characters are most clearly exhibited in the peculiarly flexuous disposition of the columella, the deflected and produced canal, obliquity of the columellar plaits, and the absence of an umbilicus. The general form of the shell, too, is rather pyriform than bucciniform.

Mazzalina bulbosa, nov. sp. Fig. 7.

Shell bulbiform, or imperfectly pear-shaped, ventricose, thin in substance; spire conical, made up of about six convex whorls; aperture about two-thirds the length of the shell, oval, and produced anteriorly into a (deflected) canal of moderate length; outer lip thin, notched below the sutural line, forming a prominent sinus, and crossed on its inner face by numerous parallel, raised (revolving) lines; columella conspicuously arched or flexed, with some five to seven oblique plaits, the lower two of which are much more prominent than the others; no umbilicus; general surface of shell smooth.

Length, two and half inches; width across the centre, somewhat more than an inch and a third.

Caloosahatchie, in the banks below Fort Thompson.

This species appears at first sight very much like the Eocene *M. pyrula*, but may be readily distinguished by its greater ventricosity, the convexity of the whorls, and the absence of the sutural carination (or more properly, sub-angulation). The columellar plaits are also considerably more prominent in the Florida form.

As far as I am aware these are the only two species that can be properly included in the genus, which appears, consequently, to be extinct. Whether or not the modern Lagena, which so closely resembles it, is immediately related, remains to be determined.

Voluta Floridana, nov. sp. Fig. 8.

Shell fusiform-ovate, smooth, except two or three of the terminal whorls of the spire, which are longitudinally ribbed, and cancellated by a number of delicate revolving lines; spire elevated, of about six volutions, terminating in a slightly papillated apex; whorls convex, hollowed above the middle, the depression in the upper whorls forming a subsutural band or carination.

Body-whorl smooth, about four-fifths the length of the shell, covered with delicate revolving lines, which become obsolete with age; aperture of nearly equal length, truncated at base; columellar surface with four prominent oblique folds, the upper of which is generally the longest.

Length of longest specimen somewhat less than six inches; greatest width, 2.3 inches.

Abundant in the banks of the Caloosahatchie below Fort Thompson.

I have little hesitation in affirming that this shell is the probable ancestor of the recent *Voluta Junonia* of Chemnitz which is occasionally obtained along the western keys (Egmont Key, etc.) and in the deeper waters of some of the inner bays. So close is the resemblance between the two species that at first sight there would appear to be not even the most insignificant characters by which to separate them. Although the fossils are in nearly all cases badly worn, yet in some the indications of color are still fairly preserved, which lead me to conclude that the general coloration of the shell was much as in the recent form. About the only character of any significance that I can indicate which might serve to distinguish the two species lies in the formation of the

apical portion of the spire, which is much more distinctly papillated in *Voluta Junonia*, and on which the longitudinal sulcation is but barely visible. Although these characters are seemingly of not much importance, they are, nevertheless, constant, and serve invariably to distinguish the one form from the other; were it not for them, I must admit that it would be very difficult, if not impossible, to separate the two. The Florida fossil appears, however, to attain a much larger size; at least, I have seen no specimens of the recent species that in any way begin to compare with it, nor have I seen any figures or descriptions of the shell that would lead me to infer that equally large specimens have ever been found. Still, the same is not impossible, and I am informed by Mr. John Ford, of this city, that, to the best of his recollection, specimens of *Voluta Junonia* of the size above indicated had been seen by him.

It is certainly an interesting circumstance to find a large volute so nearly resembling the recent species, yet slightly differing from it, in a geological formation antedating the present era by but a single period, and in a region corresponding to the habitat of the living form, especially where no traces of the latter are to be found associated with it. That the one is a modified descendant of the other we have, of course, no direct means of proving, but the inference in that direction is certainly very strong—indeed, almost irresistible. To assume that the Pliocene species should have become totally extinct before the modern era, and been then followed by a specifically new form almost absolutely its identical, as far as we are able to judge, and in a region which appears to have undergone during the common period but little alteration either in its physical or physiographical features, is barely consonant with our present evolutionary conceptions, and certainly far less plausible than the view which holds the interderivation of the two forms. The latter supposition, apart from its own abstract position, is further strengthened by the similar resemblances which bind together other members of the recent and extinct Floridian faunas.

If, however, this species is interesting as indicating a probable line of modification and descent progressionally, it is equally interesting as indicating a similar line retrospectively, or one leading up to it from a still earlier period. Thus, the species stands in about as intimate relation with the *Voluta Trenholmi* of Tuomey and Holmes, from the late Miocene or Mio-Pliocene deposits of South Carolina, as it does with the recent *Voluta Junonia*, and, indeed, might be properly considered to effect a passage between the two. About the only distinctive character separating it from the older form, likewise, as far as we are permitted to judge in the absence of color ornamentation and the animal itself, the one character separating the last from the recent species, is the less prominence of the spire in *V. Trenholmi*, and a corresponding rise in the shoulders of the body-whorl, giving the shell a less fusiform outline than in either *V. Floridana* or *V. Junonia.** But just in this character the last two species are absolutely in accord, yet, strikingly enough, the immature shells of *V. Trenholmi* and *V. Floridana* are, in this respect, undistinguishable. The fact that we have in the Miocene, Pliocene, and recent periods but a single species of the type here referred to inhabiting the (approximate) region under consideration, combined with the circumstance that over a considerable portion of this region no species of Voluta are any longer to be found, lends, I believe, conclusive evidence proving a case of true evolution and migration.

Mitra lineolata, nov. sp. Fig. 9.

Shell fusiform, gradually tapering; spire elevated, of six or seven volutions, terminating in a papillated apex, which, however, is wanting in all but the youngest specimens; whorls of spire deeply furrowed, three pseudo-sulcations on each whorl, formed or bounded by four sharply raised revolving lines or ridges, the lower of which is less prominent than the others, and constitutes a supra-sutural carination; suture slightly impressed.

Body-whorl gently convex, excavated below the suture, which is bounded inferiorly by a double carination; surface crossed by numerous (nearly equally placed) elevated revolving lines, which are not raised into ridges as on the other whorls; towards the base these lines become crowded, here and there appearing in pairs, which are separated by shallow sulcations; aperture semi-lunate, considerably more than half the length of the shell; outer lip thin.

Columellar surface nearly straight, crossed by seven oblique plaits, which rapidly diminish in prominence from above downwards.

Length, four inches; width, 1.3 inches.

Caloosahatchie, in the banks below Fort Thompson.

This species most nearly resembles *Mitra Carolinensis* of Conrad, originally described from the Miocene of Duplin Co., North Carolina (Am. Journ. Science, xxxix, p. 387; xli, p. 345, pl. ii, fig. 5), and subsequently identified by Tuomey and Holmes from the nearly equivalent deposits of South Carolina. The general characters of the two species are very nearly the same, and on a cursory inspection the Florida and Carolina fossils could readily be mistaken for one another. Closer examination, however, reveals the following points of difference, which I find to be constant for all the specimens of both species that I have had an opportunity to study: In *Mitra Carolinensis* the revolving lines on the body-whorl are much less

---

* The type specimen of this species, which has been kindly submitted to me for examination by Prof. Whitfield, of the American Museum of Natural History of New York, has a more markedly papillate apex than either the Florida fossil or *Voluta Junonia*.

prominent than in *M. lineolata*—this portion of the shell appearing nearly smooth—and towards the base, best seen on the back, they are no longer *elevated* but *impressed*, or impressed with a marginal carination, a feature not seen in *M. lineolata*, in which all the lines are elevated; in *M. Carolinensis* the infra-sutural excavation, besides being narrower, is clearly defined on several of the whorls, whereas in *M. lineolata* it is apparent as such only on the body-whorl; the whorls of the spire in *M. Carolinensis* are much more distinctly convex, whereas the elevated ridges are not nearly as sharply defined as they are in *M. lineolata*. The young of *M. lineolata* is equally ridged over the entire surface, which does not appear to be the case with *M. Carolinensis*.

Despite the differences here indicated, there can hardly be a doubt, it appears to me, that the two forms are merely derivatives one from the other—an expression of the ceaseless law of evolution.*

Marginella limatula, Conrad.

Jour. Acad. Nat. Sciences, Phila., vii, p. 140.

I identify a shell from the Caloosahatchie with this species, which may, perhaps, be properly considered to be only a variety of the recent *M. apicina* of Mencke. The typical forms of the latter, however, differ in the produced spire and in lacking the prominent denticulations on the labrum. In the series of Marginellas obtained by Gabb from the Miocene deposits of Santo Domingo, and identified by that paleontologist with *M. apicina*, we have the gradual passage leading from the high-spired form to the form in which the apex is almost completely buried in the callus developed by the rising outer lip; the character of the crenulations on the labrum is also shown to vary considerably. Gabb's form appears to be identical with the species from the Caloosahatchie, and, as far as I can determine, is undistinguishable from the Miocene fossil of the Carolinas and Virginia. The species referred to by Tuomey

---

* Since writing the above I have obtained for comparison, through the kindness of Prof. Whitfield, of the American Museum of Natural History of New York, the specimen which Tuomey and Holmes identified with Conrad's species. This shows a character of ornamentation more nearly that of the Florida fossil, but the convexity of the spiral whorls, and the lesser prominence of the revolving ribs or sulcations, serve readily to distinguish it from that species. It really stands intermediate between the typical *Mitra Carolinensis* and *Mitra lineolata*, although nearer the former, and whether all three should now be united into a single species, or the extremes, which are very well marked, be retained apart, is a matter of little import. It is manifest that with the continued discovery of intermediate forms the classification of species (and no less that of genera, etc.) will become more and more artificial and arbitrary, necessitating ultimately, if convenience is still to be considered, the placing of the same "specific" form, with its attendant varieties, into a series which might include distinct species, genera, families, or even orders. For it cannot be denied that the relationship established through phylogeny is at least as important from a classificatory point of view as that furnished by the taxonomic characters derived from living forms alone.

and Holmes as being abundant in the Post-Pliocene deposits of South Carolina, and but barely differing from *M. limatula*, is undoubtedly the recent *M. apicina*.

Oliva literata, Lamarck.
Annales du Muséum, xvi, p. 315.
Tryon, Manual of Conchology, v, p. 83, pl. 31, figs. 5-7.

Oliva reticularis, Lamarck.
Annales du Muséum, xvi, p. 314.
Tryon, Manual of Conchology, v, p. 83, pl. 30, figs. 91-95; pl. 31, figs. 96, 4; pl. 34, fig. 57.

Both of these forms, as far as I am able to judge, are represented among the fossils of the Caloosahatchie. The two species, however, so very closely resemble one another in the general characters of the shell, that I am far from certain that they can in all cases be distinguished in the absence of color-markings. The produced and more attenuated spire of *O. literata*, which may serve in the majority of instances to separate this species from *O. reticularis*, is not a constant distinguishing character, inasmuch as we sometimes find the relative condition of this portion of the shell reversed; *i. e.*, depressed in *O. literata* and elevated in *O. reticularis*. A more constant character can, perhaps, be obtained from the direction taken by the basal columellar folds, which, as a rule, are slightly more transverse and arched in *O. reticularis*—more nearly direct in *O. literata*. I must admit, however, that the correspondences and divergences seen in these minor characters give but insecure grounds for either the determination or separation of the species; indeed, it appears to me, it might be fairly questioned whether the two living forms here indicated are not in reality only varieties of one and the same species.

Columbella rusticoides, nov. sp. Fig. 9*.

Shell turreted, with an acute spire of some six volutions; whorls convex, impressed below the suture, the uppermost obscurely plicated, the lower ones indistinctly (longitudinally) lined; body-whorl high, flattened on the shoulder, and ornamented with numerous revolving lines (or bands), the upper of which are nearly obsolete; aperture ascending, narrow, somewhat more than one-half the length of the shell; outer lip thick, coarsely crenulated; columellar surface with five or six basal beads.

Length, .5 inch.

From the banks below Fort Thompson.

This species is very close to the recent (European) *C. rustica*, differing from it mainly in the upper angulation of the body-whorl and the subsutural sulcation of the whorls generally. It is a little remarkable that it should approach a trans-Atlantic form more nearly than any of the American species.

Cancellaria retioulata, L.
Syst. Nat., 12th ed., p. 1190.
Tryon, Manual of Conchology, vii, p. 69, pl. 2, figs. 25, 26.

Banks of the Caloosahatchie, below Fort Thompson.

Pleurotoma limatula ? Conrad.
Journ. Acad. Nat. Sciences, Phila., vi, p. 224, pl. ix, fig. 12.

Several individuals of a small Pleurotoma were found in the banks below Fort Thompson, which agree very closely with the Miocene fossil from Maryland, differing from it mainly, or solely, in a somewhat pronounced acuteness of the obliquely directed ribs. The limited number of specimens at my command prevent me from absolutely determining a specific identity, which, however, I firmly believe exists. The species represents *P. Suessi* from the Vienna Basin.

Conus Tryoni, nov. sp. Fig. 10.

Shell sub-conical, sinistral, rather thin in substance; spire more elevated than in the typical cones, of about eight or nine volutions, terminating in a prominent pointed apex; whorls of the spire subangulated, or carinated above the suture, the carination sharply but minutely crenulated on the first five or six whorls; suture bordered inferiorly by a prominent raised convex line, which is followed by from four to five less prominent (and occasionally quite obscure) revolving lines on the shoulders of the whorls.

Body-whorl about four-fifths the length of the shell, gently convex, crossed for the greater part of its extent by numerous obscure lines or composite bands, which become conspicuous toward the base, and exhibit there a distinct, although irregular, alternation of coarser and finer lines.

Aperture somewhat arcuate, broadest near the base; columellar surface slightly folded basally; outer lip thin; sinual inflection a half-inch in depth.

Length, five inches; greatest width, 2.3 inches.

Caloosahatchie, in the banks below Fort Thompson.

This beautiful cone, by far the largest reversed species of the genus with which I am acquainted, can readily be distinguished from the only other sinistral form that has thus far been described from the Tertiary deposits of the Eastern United States, *Conus adversarius*, by its more ponderous proportions, the greater relative elevation of the spire, and the revolving lines on the shoulders of the whorls. These last are obscured through erosion in some specimens, which then more nearly approach the Miocene fossil. There appears to be a narrower form of this type, which possibly represents a distinct species. It differs in the more pronounced angulation of the body-whorl, the lesser relative width of the crown, and a more pronounced straight-sidedness in the bounding lines of the shell. The number of such specimens in our collection is not very great, and scarcely sufficient to warrant a specific separation of the form from the species just described.

Named after Geo. W. Tryon, Jr., the distinguished conchologist of the Academy of Natural Sciences of Philadelphia, from whom the author has received much valuable assistance in the preparation of this and other paleontological papers.

Conus Mercati ? Brocchi.
Conchiologia Fossile Subapennina, ii, p. 287, pl. 2, fig. 6.
Hörnes, Die foss. Mollusken d. Tertiärbeckens von Wien, ii, p. 23.

Shell obconical, broad, straight-sided; spire moderately elevated, gradually sloping for about six volutions, then abruptly elevated in the apex; total number of volutions about twelve; surface smooth; aperture nearly straight and parallel-sided; columellar folds obscure.

Length, 2.3 inches; width, 1.3 inches.
Caloosahatchie, banks below Fort Thompson.

The species of cone here described so closely resembles in its general features *Conus Mercati* of Brocchi that I fail to find any distinguishing characters by which to separate it from that form, with which I have accordingly doubtfully united it. It must be admitted, however, that the determination of the species of Conus is a very difficult one, rendered doubly so in the absence of all color-markings. In our specimens, unfortunately, no markings remain; hence, despite the general agreement, some little uncertainty must still attach to an identification which neglects one of the primary distinguishing characters.

The species is apparently also closely related to a Santo Domingo fossil which Gabb identifies (doubtfully) with Michelotti's *C. Berghausii*, but differs from it in the more pronounced angulation of the shoulder, and the greater elevation of the apex. From the Miocene *Conus Marylandicus*, of Conrad, it can readily be distinguished by its much greater width, the comparatively depressed spire, and the absence of carinations on the whorls of the spire.

Conus catenatus ? Sowerby.
Quart. Jour. Geol. Soc. London, vi, p. 45, pl. 11, fig. 2.

I refer to this species a number of small cones which agree so closely with the Santo Domingo fossil as to be barely separable from it. The only differences that I can detect, and these are but very faintly indicated, are a slight concavity in the outline of *S. catenatus*, and a somewhat more pronounced elevation above one another of the whorls of the spire.

Strombus Leidyi, nov. sp. Fig. 11.

Shell of the general habit of *Strombus accipitrinus*, thick, ponderous, and, barring the wing, oblong-fusiform, with an abruptly reflected base; spire elevated, somewhat less than one-half the length of the shell, and consisting of from eight to ten volutions; the whorls flattened, the upper slightly nodose along their basal margins; surface of the whorls ornamented with numerous elevated revolving lines, which alternate irregularly in size—from twelve to fifteen on the larger whorls—and are crossed at right angles by longitudinal faintly-waved creases or ridges, representing lines of growth; body-whorl slightly concave on the shoulder, and projected anteriorly into a symmetrically curved wing, which ascends to about the middle of the penultimate whorl of the spire, and whose furthest expansion corresponds approximately with the centre line of the shell; body-whorl faintly tuberculated on the shoulder, the seven or more tubercles continued as so many distinct ridges extending about half-way to the base, and showing a tendency in some specimens to develop into true nodes; surface of the shoulder covered with numerous slightly-waved or crenulated concentric lines, which below the shoulder are replaced by broad regularly-placed bands (about fifteen or sixteen in number, and measuring about five to the inch), which, more especially on the expanded portion of the wing, can be clearly seen to be of a composite nature; wing very thick, thickest near the margin, and but faintly reflected; columellar surface covered with a thick deposit of callus, which extends nearly to the posterior apex of the recurved base.

Length of largest specimen, eight inches; greatest width, five inches.

Very abundant in the banks of the Caloosahatchie below Fort Thompson, where it is found from the water-line to the base of the *Venus cancellata* (Post-Pliocene) bed; I am not sure that we obtained any specimens from the latter deposit, but, doubtless, the species is also found there. Specimens despoiled of the wings have the form of Conorbis, and could readily be mistaken for giant species of that genus, especially as there is a slight sinual flexion in the lines of growth over the shoulder.

This beautiful stromb, which I take pleasure in naming after the distinguished President of the Philadelphia Academy, has unquestionably its nearest ally in the recent *Strombus accipitrinus*, Lam. (*S. costatus*, Gmelin), of the West Indian seas. Indeed, its resemblance to certain varietal forms of this species is so great, that in the absence of specimens for immediate comparison the one might almost be mistaken for the other, and I feel confident that in the recent form we have merely a derivative from the fossil; in other words, that the fossil species is the direct or immediate ancestor of the living one. This conclusion is supported, apart from the general characters uniting the two species, and the circumstance that *S. accipitrinus* is the only form now living in the region, or elsewhere, which at all approximates in structure the fossil species, by the ready adaptability to variation which the recent form exhibits. So marked, indeed, is this tendency to vary that some of the extreme varieties of the species might almost be said to approach more nearly *S. Leidyi* than their own type forms. This variation is seen in the flattening of the whorls of the spire, the less prominence of the wings, and in a reduction in the size of the tubercles of the body-whorl, characters, in the accentuation of which, primarily, *S. Leidyi* differs from *S. accipitrinus.* In the more common, or what might be called typical, forms of the latter species the wing is quadrangular, exhibiting its greatest expansion above the shoulder-line of the shell; in the greater number of individuals of *S. Leidyi,* on the contrary, the wing has a regular crescentic outline, although a tendency toward quadrangulation is very apparent in many of the specimens. It might be said that the two species vary toward each other in respect of this character, the one showing a tendency toward losing the quadrangulation of the wing, the other toward assuming it. A like variable feature separating the two species is exhibited in the nodulation (*S. accipitrinus*) or non-nodulation (*S. Leidyi*) of the whorls of the spire. Of much more permanent value as distinguishing characters are the greater elevation and flattening of the spire in *S. Leidyi,* the absence of true tubercles on the body-whorl, and the much greater ponderosity of the shell generally.

Lister figures a stromb (pl. 856), *Strombus integer* of Swainson, which in many respects, especially in the form and structure of the wing, recalls the Florida fossil. The species is described by Gray in his " Descriptive Catalogue of Shells " (June, 1832, p. 2) as follows: " Shell ventricose, solid, white; spire elongate, conical; last whorl nodulose behind; lip thick, rounded, white." Most authors, it appears, have failed to identify this species, described from a figure alone, as a member of the living fauna, and have accordingly discarded it from their catalogues. Mörch, however, claims it as a good species, and adds (Malakozoologische Blätter, xxiv, p. 17): " In 1869 I obtained by [from] Mr. Landauer, at Frankfurt, a specimen from a French collection marked ' *S. inermis,* Florides,' exactly corresponding to Lister's figure. It is the only [one] I recollect to have seen." His description of the species is as follows: *Testa planiuscula, solidula, albescens; spira elongata, acuta, conica; ultimo anfractu postice leviter noduloso; labro tenui expanso.* Whether or not Mörch and Gray refer in their descriptions to the same species, it is a little difficult to determine, despite the assurance given us by the former that his specimen corresponds absolutely with Lister's figures. There can be little question from Lister's drawing that the specimen intended to be represented by him has a thick lip, as correctly interpreted by Gray, whereas Mörch maintains that the lip is thin. The general form of

Lister's species, especially the outline of the wing, is so unlike that of any stromb, except the Florida fossil, that one might be readily tempted, making due allowance for imperfect drawing, to unite the two into a single species, the more especially as the shell is described (by Gray) as being white, a feature foreign to the recent representatives of the family, and indicative to a certain extent of a fossil condition. Indeed, the only marked difference between the two forms appears to lie in the more pronounced nodulation of the spire in *S. integer*, a feature not unlikely exaggerated in the drawing, which is manifestly erroneous in the delineation of the spire. Were it not for Mörch's positive statement that he has secured a specimen, with a thin lip, absolutely conforming to Lister's drawing, I should have felt little hesitancy in relegating Swainson's species to the category of fossils, and of uniting with it the species from the Caloosahatchie; and even now I am far from convinced that this identity does not exist, but as it appears practically hopeless to positively identify any form with the *S. integer*, I have deemed it the safer plan to describe the Florida fossil as a distinct species.

Strombus pugilis, L.
  Linnæus, Syst. Naturæ, 12th ed., p. 1209.
  Tryon, Manual of Conchology, Part xxvi, p. 109, pl. 2, figs. 13-15.

This species, in the variety known as *S. alatus*, Gmel., is fairly abundant in the *Venus cancellata* bed at Fort Thompson, but less so in the underlying Pliocene deposits. The specimens obtained do not differ essentially from the recent form.

Genus CYPRÆA.

Subgenus Siphocypræa, Heilprin.

I propose this subgenus for a group of remarkable Cypræas, which differ from all other members of the family in the possession of a deep, comma-shaped sulcus or depression, occupying the apical portion of the shell, and which, as the posterior continuation of the aperture, is curved dextrally around the axis of involution. It would appear that the presence of this sulcus is due to a siphonal prolongation of the mantle, which, contrary to what is seen in other Siphonata, must have been projected in advance of the animal; otherwise, the position of the sulcus would have been posterior, instead of anterior, to the apical axis. The other characters of the shell are those of Cypræa generally. In the absence of positive knowledge respecting the organization of the animal, I have retained it provisionally under Cypræa, although not improbably the distinguishing characters above indicated are of generic value.

Cypræa (Siphocypræa) problematica, nov. sp. Fig. 12.

Shell ovately cylindrical, completely involute, exhibiting in the apical region a deep comma-shaped depression—prolongation of the aperture—which is wound dextrally around the axis of involution.

Base plano-convex, slightly tumid superiorly; aperture somewhat eccentric, narrow, arcuate, continued into the comma-shaped depression above referred to; canal short, broadly reflected; labrum with from twenty to twenty-five prominent plaits, which are considerably stronger, and much less crowded, than the equally numerous plaits on the columellar surface.

Length, 2.7 inches; width, 1.4 inches.

Common in the banks of the Caloosahatchie below Fort Thompson.

This species is a much narrower shell than the Miocene *Cypræa Carolinensis*, and differs in like respect from all the larger Tertiary species of the West India Islands with which I am acquainted. It most nearly approaches in outline the recent *Cypræa exanthema*, of the Florida coast. The remarkable comma-shaped depression on the apical portion serves to distinguish it readily from all other species of the genus, either recent or fossil, that have come under my notice.

Pyrula reticulata? Lam.
    Animaux s. Vertèbres (Ficula), ix, p. 510.
    Tryon, Manual of Conchology, vii, p. 265, pl. 5, fig. 28; pl. 6, fig. 33.

Caloosahatchie, in the banks below Fort Thompson.

Mr. Tryon separates the common species of the southern United States (*P. papyratia* of Say) from the eastern *P. reticulata*, observing that the shell of the former is slimmer and more delicately sculptured. Whether these seemingly trivial characters are constant or not in the living forms I am not prepared to say, but, obviously, the ornamentation of the Florida fossil more nearly resembles that of Lamarck's species than of the presumably distinct form described by Say.

Natica canrena, L.
    Mus. Ulr., p. 674.
    Tryon, Manual of Conchology, viii, p. 20, pl. 4, fig. 58.

Caloosahatchie, banks below Fort Thompson.

Natica duplicata, Say.
    Journ. Acad. Nat. Sciences, Phila., ii, p. 247.
    Tryon, Manual of Conchology, viii, p. 33, pl. 12, fig. 3.

Below Fort Thompson.

Crucibulum verrucosum, Reeve.
    Conch. Icon., ii, Crucibulum, Species 19.
    Tryon, Manual of Conchology, viii, p. 119.

Turritella perattenuata, nov. sp. Fig. 13.

Shell very slender, gradually tapering; whorls very numerous, doubly carinated, the carinæ crenulated or beaded, the upper and lower about equally removed from the upper and basal margins of the whorls respectively, the upper carina frequently appearing double through the presence of a contiguous additional line; shoulder of the whorls prominent, with one or two elevated lines; the concave space between the carinæ with two obsoletely crenulated lines, the upper of which is somewhat the more prominent. Aperture quadrangular.

Length of a restored specimen nearly five inches; greatest width, .6 inch.

Common in the banks of the Caloosahatchie below Fort Thompson.

This shell can be at once distinguished by its extremely elongated or attenuated outline, surpassing in this character all other forms of the genus with which I am acquainted, either recent or fossil. It bears a (superficially) close resemblance to *Turritella tornata* of Guppy (Q. Journ. Geol. Soc., London, xxii, p. 580), a Miocene fossil of the island of Santo Domingo, but differs in its more slender outline, the greater relative elevation of the shoulder, and in the less prominence of the two intermediate lines between the carinæ. These are also much more distinctly beaded in *T. tornata*. Gabb maintains ("Santo Domingo," Trans. Am. Philos. Soc., xv, p. 240) that Guppy's description applies only to a single variety of the species, and enumerates other characters which are by him held to cover other varieties of the species as well. I fail, however, to see upon what ground this emendation to the original description is made. The specimens in Mr. Gabb's collection marked *T. tornata* certainly do embrace two or more distinct forms of Turritella one of which is indisputably Guppy's species, but why these should be all linked together as a single species I do not exactly comprehend. It is true that they bear a general resemblance toward one another, both in outline and ornamentation, but I fail to detect any gradual passage of the one form into the other—a condition which might naturally be expected on the hypothesis of specific identity—at any rate, not into the form which accords precisely with Guppy's description.

Turritella apicalis, nov. sp. Fig. 14.

Shell gracefully tapering, with an acute apex; whorls numerous, straight-sided, carinated above and below, the carinæ about equally removed from the upper and lower sutures respectively, distinctly beaded; a prominent subsutural line, placed about medially on the shoulders of the whorls; the flattened space between the carinæ with an obscurely beaded sub-median line, and numerous finer lines, which are almost invisible to the naked eye; the beads of the carinæ oblique, and inclined in opposite directions, the upper ones downward to the right, the lower ones downward to the left; aperture quadrangular.

Length, 1.7 inches; width of base, .3 inch.

Caloosahatchie, abundant in the banks below Fort Thompson.

Turritella cingulata, nov. sp. Fig. 15.

Shell elevated, straight-sided; whorls flat, faintly carinated inferiorly by an obscurely beaded (or "roped") line or band, which is followed successively in the direction of the apex by two distinctly crenulated or beaded lines, an obliquely and obscurely lined (barely elevated) band, and a delicate terminal line; the band above the two lines is more distinctly beaded along its base, appearing somewhat like a third crenulated line; aperture quadrangular.

Length, 2.4 inches; width of base, .5 inch.

From the banks below Fort Thompson.

Turritella mediosulcata, nov. sp. Fig. 16.

Shell rapidly tapering, straight-sided; whorls flattened, appearing somewhat concave through the presence of a depressed median area or band, which is bounded inferiorly by a fairly prominent beaded line; surface covered with numerous fine revolving lines, which above the medial depression are cut obliquely (downward to the right) by obscure rugations; aperture quadrangular; base flat.

Length of fragment, 1.5 inches; width of base, .4 inch.

A solitary specimen, from below Fort Thompson.

Turritella subannulata, nov. sp. Fig. 17.

Shell turreted, acuminate; whorls angular, marked by a broad basal impressed band or channel, which is ornamented with numerous delicate revolving lines; surface of the whorls above the channel longitudinally plicated, with two well-defined submedial lines, and numerous finer lines, as in the channel; aperture rounded; base convex.

Length, 1.2 inches; width of base, .25 inch.

Abundant in the banks below Fort Thompson.

Cerithium atratum ? Born.

Mus. Cæs., p. 324, pl. ii, figs. 17, 18.

I doubtfully refer to this species a solitary specimen found in the banks below Fort Thompson. It is a somewhat more elevated shell than the recent form, and its ornamentation also differs slightly; but on the whole its facies is very similar, and I am inclined to believe that among a selection of specimens individuals would be found to grade into the typical *C. atratum*, which is itself a markedly variable species. The specimen measures an inch and a half in length.

Cerithium ornatissimum, nov. sp. Fig. 18.

Shell acuminate, gracefully tapering; whorls numerous, fifteen or more, furrowed below the suture, and rugated with a very elaborate ornamentation; the upper seven or eight whorls of the spire distinctly plicated (longitudinally), the plications on the lower whorls becoming obsolete, and replaced by broken nodes, which are disposed in a double series —one row above the sub-sutural furrow, the other immediately below it —the nodes of the two series at first opposite, then alternate; the lower portions of the whorls granulated; body-whorl with four distinct lines of granulations, the basal one, which is separated by an interval from the others and followed by three elevated, non-granulated lines, the strongest; entire surface of the shell covered with fine revolving lines, which alternate in size; aperture about one-fifth the length of shell, gently arcuate.

Length, somewhat above two inches; width, a half-inch.

A solitary specimen from the banks below Fort Thompson.

This species can be readily distinguished by its form and ornamentation from the recent *C. atratum* and *C. eburneum*, to both of which it bears a general resemblance.

LAMELLIBRANCHIATA.

Panopæa Menardi, Deshayes. Fig. 19.
    Dict. Class. d'Hist. Nat., xiii, p. 22.
    Panopæa Faujasi (auct.).

Several large specimens from the banks below Fort Thompson, with the valves still attached.

I can find no characters by which to separate the Florida fossil from the well-known species of the European Miocene and Pliocene formations. It is almost without doubt the species figured and described by Say as *Panopæa reflexa* (Journ. Acad. Nat. Sciences Phila., vol. 4, p. 153, pl. xiii, fig. 4), which is stated to have the "shell transversely oblong-subovate; anterior margin somewhat narrower and longer than the posterior margin, the edge reflected; surface wrinkled, and profoundly so towards the base. Length, three inches and two-fifths; breadth, five inches and seven-tenths." In the above description *posterior* [margin] should stand for *anterior* and *vice versa*; *height* for *length* and *length* for *breadth*. I have not seen any specimens from the American Miocene deposits which correspond with Say's figures. The form that has been identified with it by Conrad—of which the collections of the Academy possess numerous specimens—and which has been generally accepted as Say's species, is a very different shell, easily recognized by its declining posterior slope, the position of the umbones, which are almost invariably placed nearer the posterior border than the anterior, and the acute angulation and narrowness of the sinual inflection. In Say's *P. reflexa*, as well as in the Florida fossil and *P. Menardi*, the hinge-line is about equally elevated both anteriorly and posteriorly, the beaks are somewhat anterior, and the sinual inflexion relatively shorter and broader. The shell is also more massive. In view of the differences here indicated, I would propose for the common form of the Atlantic Middle Tertiaries, hitherto referred to *Panopæa reflexa*, the name of *P. cymbula*, the species to be defined as follows:

Panopæa cymbula, nov. sp. Fig. 20.
Panopæa reflexa, of most authors.

Shell expanding anteriorly, where it is highest, gracefully rounded; rapidly sloping posteriorly, with the border reflected, permitting of a broad gape; beaks somewhat posterior, or beyond the middle, considerably sloping, the apex directed slightly to the rear; a prominent transverse cardinal tooth beneath the apex, followed by a strongly-bordered cartilage plate; muscular and pallial impressions rugged, deep; sinual inflection generally narrow and acutely pointed; external surface of shell strongly and roughly furrowed.

Length, 5.3 inches; height, three inches.

Miocene of the Atlantic slope.

Panopæa Floridana, nov. sp. Fig. 21.

Shell oblique, expanding and ascending anteriorly, abruptly truncated behind; hinge-line in front of the umbones rising considerably, declivous beyond the cartilage-plate, and ascending again toward the posterior extremity; posterior margin reflected, the shell gaping broadly; umbones well in front of the middle; cartilage-plate very strong; ligamental sulcus deep; muscular and pallial impressions well impressed, rough, the sinual inflection often v-shaped.

Length, 5.1 inches; height, three inches.

Both valves of a single individual.

This species can be readily distinguished from *P. Menardi* (*P. reflexa*, Say) by its truncated form, and the rise in the hinge-line in front of the umbones; the height of the gape is also relatively greater.

Panopæa navicula, nov. sp. Fig. 22.

Shell (known only by the right valve) short, broadly-oval, obliquely rounded anteriorly, abruptly truncate behind, the gape (posterior) very broad; umbo in advance of the middle of the shell, the apex directed forward; hinge-line sigmoidal, or flexuously curved, ascending in front, reflected posteriorly; cardinal tooth prominent, arched upward; cartilage plate strong; muscular and pallial impressions very deep, the sinual inflection short and openly quadrangular; surface of shell prominently sulcated.

Length, five inches; height, 3.5 inches.

This shell most nearly approaches the preceding in outline, but can be distinguished by its broadly swelling anterior border, its greater height, and the form of the sinual inflection.

Semele perlamellosa, nov. sp. Fig. 23.

Shell thin, transversely elongated, oval, about equilateral; beaks central, acute, not prominent; right valve with a vertically incised, lamellar, cardinal-tooth, which is followed by the very oblique, and broadly opening cartilage sulcus; dental fissure in left valve narrow, apparently duplicated; lateral teeth subcentral; surface covered with numerous, regularly placed, elevated lamellæ of growth, which are gently angulated on the posterior slope.

Length, 2.8 inches; height, two inches.

A single specimen from the banks below Fort Thompson.

Venus rugatina, nov. sp. Fig. 24.

Shell ventricose, subcircular, appearing on its inner aspect much like Isocardia, with a prominent anterior projection, corresponding to the lower boundary of the deeply-impressed lunule; hinge-line in each valve with two prominent, transversely projecting, cardinal teeth, and a third (posterior) oblique one, which, in the left valve, is separated from the median tooth by a deep, obliquely directed, dental sulcus; the anterior tooth in the left valve deeply grooved above, and preceded by a dental papilla; posterior lateral teeth not prominent; surface covered with very numerous, sharply-defined, imbricated lines of growth, which at nearly regular intervals are marked off by lines of special prominence, immediately below which the normal lines are least closely crowded; number of prominent lines twenty-five and upward; number of normal lines between these from six to eight; border crenulated, the crenulation extending under the lunule.

Length, 2.5 inches; height, 2.1 inches.

Abundant in the banks below Fort Thompson.

Closely resembles the recent *Venus rugosa*, but may be readily distinguished by the greater interval separating the more prominent lines.

Cardium Floridanum, nov. sp. Fig. 25.

Shell obliquely cordate, upright, posterior slope somewhat depressed, flattened; beaks elevated, with the apex turned gently toward the anterior margin; left valve with a deep pit immediately under the apex, in front of which is a prominently projecting pointed tooth; two pyramidal lamellar teeth, underneath which are corresponding dental pits; right valve with the lateral teeth underlying the dental pits; hinge-line raised anteriorly into a flattened vertical plate, which ascends beyond the apex of the beak.

Surface radiately ribbed; ribs narrow, about thirty in number, profoundly squamous, the scales erectly vaulted, compressed and carinate, or overlapping, on the posterior side, broadly flattened anteriorly, so as to produce shallow cups, through which the ribs appear to pass.

Height, 1.4 inches.

This species most resembles among recent forms *Cardium consors* from the west coast of South America, which differs in the greater number and closer imbrication of the scales on the posterior slope, and in lacking the open cup-like forms of the anterior border; the elevated dental plate is also less prominent.

Hemicardium columba, nov. sp. Fig. 26.

Shell (known only by the left valve) elevated, gibbous, carinated on the umbonal slope; posterior cordiform space profoundly hollowed; anterior border evenly rounded; base slightly sinuous posteriorly, somewhat produced; posterior border deeply emarginate; beaks elevated, transverse, the apex appearing as though directed to the rear; cardinal teeth two, enclosing a deep pit, the anterior one much the more prominent; lateral teeth lamellar, pyramidal; entire surface of shell closely ribbed, ribs echinated, about thirty-six in number, some thirteen of which are on the posterior slope; basal margin crenulated.

Height, to summit of beak, four-fifths of an inch; length, .7 inch.

I have but two valves of this species, a near ally of the recent *Hemicardium media* of the southern coast, from which it can be distinguished by its more upright form, the deeper hollowing of the posterior face, and the more pronounced carination of the umbonal slope.

Chama arcinella, L.

Syst. Nat., p. 1139.

Numerous in the banks below Fort Thompson.

The individuals of this species vary in the fossil much as they do in the recent form, the variation depending upon the convexity of the valves, the disposition and thickness of the spines, the presence or absence of interstitial secondary radiating lines, etc. In the collections of the Academy of Natural Sciences there are a number of Chama valves, from Cape Fear River, North Carolina, which are marked in Conrad's handwriting *Arcinella spinosa*. I am not aware that the species has ever been described as such, but it is nothing other than an unusually spiny variety of *C. arcinella*, and, doubtless, the form which is correctly referred in Emmons's North Carolina report to the living species.

Chama crassa, nov. sp. Fig. 27.

Shell thick, ponderous, cordiform, with a prominent sulcus descending the posterior slope; both valves very convex, the left one somewhat the larger; beaks spirally twisted, much as in *Isocardia*; the hinge-line (in the left valve) with a profound arciform ligamental furrow, and two nearly horizontally placed cardinal teeth, separated by a broad, slightly curved, dental sulcus; muscular impressions sharply defined, deep; external surface rugose, covered with the somewhat sinuous lines of growth; basal margin entire, not crenulated.

Length of largest specimen, measured along the antero-posterior axis, and through the umbones, 3.7 inches; height, nearly three inches; depth of a single valve somewhat over two inches.

Common in the banks below Fort Thompson.

This form may be readily distinguished from all the other species of North American Chama, either recent or fossil, by its ponderous form, and the near equality of the two valves, differing in the latter character conspicuously from the European *Chama gryphoides*, which much resembles it. It differs in this respect also from the American *Chama corticosa* of Conrad (Miocene), which, however, is a sinistral shell.

Lucina disciformis, nov. sp. Fig. 28.

Shell compressed, suborbicular, higher than broad, flattened on the umbonal slope; beak subcentral, acute, overlooking a deeply impressed lunule; ligamental sulcus profound; cartilage-pit oblique; cardinal teeth two in each valve, the posterior in the right valve, and the anterior in the left valve, bifid; anterior margin of shell impressed somewhat above the middle; muscular impressions elevated, the anterior ribbon-form, long and narrow, departing somewhat from the rather distantly separated pallial line; external surface ornamented with numerous distantly placed lines of growth, which at nearly equal intervals rise into rugose elevated lamellæ; interior of shell longitudinally rugated.

Height, 2.5 inches; width, 2.4 inches.

This shell bears a similarity of outline to the Miocene *Lucina Americana* (*L. anodonta*), and is barely distinguishable from that species by external characters alone; the latter is, however, edentulous. Among recent forms it approximates *L. Childreni*, but that species is inequivalve.

Area scalarina, nov. sp. Fig. 29.

Shell obliquely rhomboidal, elevated, ventricose, angulated posteriorly, flattened; anterior end short, evenly rounded; beaks prominent, transverse, about eight, distant; ligament-area diamond-shaped, nearly smooth in the young shell, with delicate transverse lines—in the adult, with a limited number of coarse, sinuous longitudinal lines; hinge-line straight, somewhat more than one-half the greatest length of shell; teeth numerous, somewhat oblique toward either end.

Ribs prominent, about twenty-four, broad, square, robustly crenate, those of the left valve broader than the interspaces, flattened posteriorly, about eight on the anal angulation; those of the right valve of about the same width as the interspaces (the anterior ones the broadest), with an interstitial secondary rounded rib in the centre of the interspace; the two valves unequal, the basal margin of the left valve greatly protruding beyond that of the right; base profoundly crenulated.

Length, 3.3 inches; height, 2.5 inches.

Abundant in the banks of the Caloosahatchie below Fort Thompson.

I am not absolutely satisfied as to the value of this species, although the form, so far as all the specimens collected by us are concerned, is a very clearly defined one. It closely resembles the shell identified by Tuomey and Holmes with *Arca scalaris* of Conrad (Foss. Med. Tert. Form. U. S., p. 59, pl. 31, fig. 1; Tuomey and Holmes, Pliocene Fossils of South Carolina, p. 43, pl. xvi, figs. 1, 2), and might, indeed, be readily mistaken for it. Through the kindness of Prof. Whitfield I have been permitted to make a comparison with the type-forms described and figured by Tuomey and Holmes, and find that their shell differs very materially from the Florida fossil. In the first place it is decidedly more oblique, and secondly, the ribs adjoining the posterior slope (on the left valve) are not nearly as broad relatively, nor as flattened, as they are in *A. scalarina;* the ribs of the left valve are more remotely placed from one another, and lack the pronounced interstitial secondary rib, which is so prominently defined in the Florida fossil. Its place is taken by a hair line, which is present in some of the intercostal spaces. The characters of the Florida shell are remarkably constant, showing practically no variation, and were I as positive of the stability of characters in the Carolina fossil, I should have no hesitation in regarding the two as specifically distinct; unfortunately, only a single pair of valves of Tuomey and Holmes's shell has been positively identified, which, therefore, gives no information on this point. As it is, the characters of the two are sufficiently distinct, indeed, fully as well-marked as those which separate the Florida fossil from the recent *Arca incongrua* of the Southern coast, which may, with much plausibility, be looked upon as its immediate descendant. The recent species agrees more nearly in the general outline of the shell, being upright rather than oblique, but differs in the less width (in the left valve) of the ribs, and in lacking the true interstitial rib of the right valve (although an indication of it appears in a faint elevated line), agreeing in this respect with the South Carolina fossil. That the three forms are most intimately related there can be no question, and I believe there is likewise little or no question that they all lie on the same line of descent. Tuomey and Holmes assert that their shell is closely allied to *Arca transversa*. This is certainly a mistake; the two shells, beyond the general characters uniting the majority of Arcas, have very little in common—neither in shape, size, nor ornamentation.

Another point that remains to be determined in this connection is whether the fossil described by Tuomey and Holmes is really the *Arca scalaris* of Conrad, or not. That paleontologist was acquainted with the left valve only of his species, nor do the collections of the Academy of Natural Sciences, which embrace the greater number of the specimens either obtained or described by Conrad, show any other portion of that species but the left valve. Inasmuch as one of the more important distinguishing characters lies in the opposite valve, it is impossible to say whether the form in question would approximate more nearly the South Carolina fossil or *Arca scalarina*, or, indeed, whether it is the equivalent of either the one or the other. All the valves, which include the figured form, are of comparatively small size, and, as far as mere outline is concerned, more nearly resemble *A. scalarina*. The posterior ribs, however, as in the South Carolina shell, are narrower than in the Florida fossil, lacking the peculiar flattening of that species, and, reasoning from the fact that the shell was obtained from the same series of deposits which also yielded the South Carolina fossil, it may perhaps be assumed that the two are identical. This is, however, a matter of conjecture. The umbones in Conrad's shell are considerably less prominent than in either the South Carolina or the Florida fossil, and possibly the form never attained a size comparable with that of either of the two other forms.

Arca crassicosta, nov. sp. Fig. 30.

Shell subquadrangular; ventricose, ponderous, ornamented with about twenty coarse, elevated, transversely barred, terete ribs, which are somewhat irregular and crowded on the anterior half of the shell, becoming widely separated and profoundly elevated on the posterior half; beaks nearly anterior, looking forward, separated from each other by a fairly broad interval; hinge-line almost as long as the greatest length of the shell, pectinated with numerous narrow, nearly vertical teeth; ligamental area narrow, elongated, with about six longitudinal lines, which rise toward the apex of the beak; interior of the shell coarsely rugated; base ascending anteriorly, profoundly crenated.

Length, 2.7 inches; height, two inches.

Below Fort Thompson.

This shell can be readily distinguished by its profoundly elevated and widely separated ribs, being the most coarsely costated Arca with which I am acquainted. It closely resembles *Arca rustica*, of Tuomey and Holmes (Pliocene Fossils of South Carolina, p. 39, pl. xv, fig. 1), and may, indeed, be that shell. Unfortunately, these authors have given but a very meagre and unsatisfactory description of their species, which barely permits of an absolute determination being made. Their figure, drawn from a mere fragment, does not represent the profound ribs seen in the Florida fossil; the posterior interspaces are apparently also much narrower, nor does there appear to be any marked variation either in the disposition or the size of the costæ. Still, the differences here indicated, which are based upon figure and description only, may be more apparent than real, and the two forms, as above intimated, may in reality represent a single species. Unfortunately for the determination of this point, the only perfect valve possessed by Tuomey and Holmes was lost before the publication of their monograph, and the fragment which served as the type of the species, for both figure and description, and which, as far as I am aware, was the only other specimen extant illustrating the species, has since been lost from the collection of the South Carolina geologists. I am informed to this effect by Prof. Whitfield, of the American Museum of Natural History, of New York city, where the type collections of the South Carolina Survey are deposited.

Arca lienosa, Say.
    American Conchology, pl. 36, fig. 1.
    Arca Floridana (recent), Conrad.

This shell is identical with the recent *Arca Floridana*, from the Florida coast, the specific name of which will have to be replaced by that of Say's species, which has priority. The only difference that it presents, and this is probably not constant, and at most a trifling character, is a somewhat greater anterior projection of the hinge-line, making the shell appear more distinctly eared. It does not appear, however, that the recent form attains the size of that seen in the fossil; one of the specimens from the upper Caloosahatchie measures six inches in length, and three inches in height to the top of the umbones. The ribs where worn, especially towards the base, show a double or quadruple structure, best seen in the larger specimens.

Arca aquila, nov. sp. Fig. 31.

Shell (known only by its left valve) elongated, rectangular, winged, profoundly sulcated on the posterior slope; anterior border vertical, straight; basal line slightly sinuous beyond the middle; posterior border deeply emarginated; hinge-line straight, of nearly equal length with the base; teeth very numerous, gradually increasing in size towards either end, where they are markedly oblique; hinge-area broad, obscurely furrowed in longitudinal lines; beak moderately elevated, incurved, the apex directed backward; surface of the shell radiately ribbed, the ribs sinuous, beaded—especially on the anterior portion of the shell, where they are separated by an intermediate fine line—becoming obsolete in the posterior sulcus and on the wing, where they are represented by two pairs of lines; lines of growth prominent towards the base and on the wing; basal margin crenulated.

Length, 1.25 inches; height, from base to hinge-line, .5 inch.
This winged ark is at once distinguished from *A. aviculæformis* (*v. post.*) by the absence of the anterior rostrum and its rectangular form. The last character, in addition to differences in the ornamentation, also serves to distinguish it from the Miocene *Arca incile*, which resembles it somewhat in the pterination of the posterior slope.

ARCA.

Subgenus Arooptera, Heilprin.

Shell elongated, aviculæform, rostrated anteriorly, winged posteriorly; hinge-line practically the entire length of the shell, exceedingly narrow, and pectinated with a crowded series of transversely directed or partially v-shaped teeth; umbones widely separated; ligamental area very broad, open; base sinuous, with a median opening.

I propose this subgenus for a series of very remarkable arks, which can be readily distinguished from all other members of the genus by their peculiar pterination and rostration, giving an external outline precisely that of Avicula. This character is accompanied by a most extraordinary attenuation anteriorly of the chamber of the shell.

Arca (Aroöptera) aviculæformis, nov. sp. Fig. 32.

Shell elongated, aviculæform, rostrated anteriorly, winged posteriorly, with a prominent obtuse carination on the umbonal slope bounding the wing; rostrum declivous, marked off from the body of the shell by a broad hollow; basal margin of shell sinuous, showing a median opening, and rapidly sloping upward in the direction of the rostrum; posterior border deeply emarginate.

Umbones acute, very excentric, moderately elevated, and but slightly incurved, with a gradual continuous slope to either extremity of shell; hinge-line nearly the whole length of shell, very narrow, pectinated with a crowded series of lamellar, transversely directed, teeth, which exhibit a tendency to become oblique and v-shaped on the posterior half of the line; ligamental area broad, open, arching upward in a gentle curve, longitudinally lined, and irregularly grooved by numerous diagonal or v-shaped furrows resembling insect borings.

Surface of shell ornamented with numerous radiating wavy lines, alternating in coarseness, which become more or less obsolete on the umbonal slope, and are wholly wanting on both the beak and wing, which only show concentric lines of growth; of the radiating lines on the anterior part of the shell the series runs about as follows: coarse line, followed by two finer lines, then a slightly more prominent single line, again two finer lines, and then a coarse line, same as first, marking the coarse lines at intervals of about six or seven; interior of shell deep, cuneiform; margin entire.

Length, 5.4 inches; width across the beaks, 2.5 inches.

Caloosahatchie, in the banks below Fort Thompson.

Spondylus rotundatus, nov. sp. Fig. 33.

Shell (known only by the larger valve) capacious, orbicular below the hinge-line, distinctly auriculated; hinge-line triangular, pyramidal, the beak acute, laterally twisted at the apex, traversed by a median slit; cartilage-pit profound, reaching about half-way to the apex; cardinal teeth powerful, slightly spreading; external surface coarsely ornamented with irregular squamous ribs and intermediate scaly fine lines, the (imbricated) scales on the latter drawn out into flattened spines or echinations.

Height to apex, 3.5 inches; greatest width, 2.5 inches.

A solitary, perfectly preserved valve from the banks below Fort Thompson.

This species is quite distinct from any form, either recent or fossil, with which I am acquainted.

Pecten solarioides, nov. sp. Fig. 34.

Shell suborbicular, depressed, of about equal height and width; ribs about 20 (?), quadrangular or flattened, broader than the interspaces, crossed by fine rugose lines of growth; a faint median longitudinal line or carination can be detected on some of the ribs, probably eroded on the others; the interspaces with two or more elevated longitudinal lines; left ear of right valve with about five very oblique, narrow ribs, rugose with the lines of growth; right wing? Interior of shell prominently ribbed.

Height, 5.7 inches.

The half of a single right valve, and a fragment of probably the other valve, from the banks below Fort Thompson.

This shell can be readily distinguished from the only species that at all resembles it, *Pecten comparilis*, by its more elevated form, the height of its wings, and the structure and disposition of its ribs, which are more distinctly quadrangular and elevated. In the fragment which possibly represents the left valve the ribs are broader than in the opposite valve, of about twice the width of the interspaces, which, at least in the upper portion of the shell, are deep and nearly parallel-sided. They show a single median elevated line.

Pecten comparilis, Tuomey and Holmes.

Pliocene Fossils of South Carolina, p. 29, pl. xi, figs. 6–10.

Pecten eboreus of Conrad (in part).

I refer to this species a number of large Pectens, found in the banks below Fort Thompson, and also at Thorpe's, some of whose forms are absolutely undistinguishable from the Carolina fossil.* The largest perfect specimen measures 4.2 inches in height, and nearly five inches in greatest width. The form appears to be a rather variable one, the variation depending upon the relative convexity of the valves and ribs, the latter, in the larger individuals, more generally assuming the flattened form, with a submarginal carination; the interspaces between the ribs may also exhibit two or more faint carinations (imbricated longitudinal lines), a character which was used by Conrad to distinguish *P. Edgecomensis* (Proc. Acad. Nat. Sciences, 1862, p. 291) from *P. eboreus*. I have not seen any specimens of *P. Edgecomensis*, and am therefore unable to say whether it corresponds to the variety of *P. comparilis* here indicated, or not.

Pecten (Pleuronectes) Mortoni, Ravenel.
Proc. Acad. Nat. Sci. Phila., ii, p. 96.

Identified by one nearly perfect specimen and several fragments from among the fossils of the upper Caloosahatchie. This species is most intimately related to *Pleuronectes Japonicus*, of the East Asiatic seas, and can only be distinguished from it by the characters drawn from the radiating raised lines of the interior of the shell, which in *P. Mortoni* are disposed in narrow pairs, passing as such to the border. In *P. Japonicus* the lines, while starting in pairs, lose their dual character long before reaching the margin.

Pecten nodosus, L.
Syst. Nat., 12th ed., p. 1145.

A solitary fragment, absolutely undistinguishable from the recent species.

Caloosahatchie, below Fort Thompson.

Ostrea meridionalis, nov. sp. Fig. 35.

Shell (at first) strongly plicated, suborbicular or elongated, becoming massive and ponderous when full-grown, with an almost complete obliteration of the plications; the plications in the adult not visible on the

---

* I have examined what is supposed to be the type-specimen of Tuomey and Holmes's *Pecten comparilis*, and find that it differs somewhat from what has generally been assumed to be species in question. The characters in which it varies, as the greater convexity of the apical portion of the shell, and a slight reduction in the number of ribs, are, however, insignificant, and inasmuch as the specimen is a solitary one, and consequently throws no light upon individual variation, I think that the identity of the species with the form that Conrad has recognized as *P. comparilis*, a shell attaining a very much larger size than that figured and described by Tuomey and Holmes, may be fairly assumed. This form, moreover, is that which was also subsequently identified with their own species by the South Carolina geologists. Whether the species is distinct from Conrad's *Pecten eboreus* is a question which, in the absence of a sufficiently large number of specimens for comparison, I am not prepared to answer.

interior faces of the valves; hinge-area, in the adult, greatly elongated, either straight or laterally twisted, sometimes nearly one-half the length of shell; the median groove of variable width and depth, projecting below the lateral ridges, so as to make the hollow of the shell appear two-horned; cavity of shell shallow, impressed medially just below the hinge-line; muscular scar semilunar, deep.

Length (height), 6.3 inches; width, variable; thickness of shell, two to two and a half inches and upward.

Very abundant in the "marl" banks below Thorpe's, where the shell can be seen several feet below the surface of the water; less abundant in the banks below Fort Thompson, and at one or two points between this locality and Thorpe's.

This form may be readily distinguished from all the other Eastern species of Ostrea by its ponderous proportions, greatly surpassing in this respect *O. percrassa* from the Miocene of New Jersey. In its early stage it resembles in both outline and ornamentation the recent *O. borealis*, but the adult form, from its manner of growth, is wholly different.

The following additional species of Mollusca were obtained from the deposits below Fort Thompson:—

Nassa vibex, Say.
    Journ. Acad. Nat. Sciences Phila., ii, p. 231.
Crepidula cymbæformis, Conr.
    Proc. Acad. Nat. Sciences Phila., ii, p. 173.
Crepidula fornicata, L.
    Syst. Nat., p. 1257.
Bulla striata, Brug.
    Dict., No. 3.
Siliqua bidentata, Spengler.
    Skrivt. Nat. Selsk., iii, p. 104.
Semele variegatum, Lam.
    Anim. s. Vertèbr., vi, p. 126.
Rangia cyrenoides, Des M.
    Actes Soc. Linn. Bordeaux, v, p. 57.
Venus cancellata, L.
    Syst. Nat., p. 1130.
Venus Rileyi, Conr.
    Foss. Med. Tert. U. S., p. 9, vi, fig. 1.
Venus Mortoni, Conr.
    Journ. Acad. Nat. Sciences Phila., vii, p. 251.
Artemis discus, Reeve.
    Conch. Icon., *Artemis*, 9.
Artemis elegans, Conr.
    Foss. Med. Tert. U. S., p. 30.
Dione (Calliste) gigantea, Chem.
    Conch. Cab., x, p. 354.

Dione maculata, L.
  Syst. Nat., p. 432.
Cardium magnum, Born.
  Mus. Cæs., pl. 3, fig. 5.
Cardium isocardia, L.
  Syst. Nat. (Gmelin ed.), p. 3249.
Lucina edentula, L.
  Mus. Ulr., 74.
Lucina Pennsylvanica, L.
  Syst. Nat., p. 1134.
Lucina Floridana, Conr.
  Amer. Journ. Science, xxiii, p. 344 (1833).
Lucina tigerina, L.
  Syst. Nat., p. 1133.
Carditamera arata, Conr.
  Foss. Shells Tert. Form. U. S., p. 20.
Arca plicatura, Conr. (et **Arca improcera**).
  Foss. Med. Tert. U. S., p. 61.
  Heilprin, Proc. Acad. Nat. Sciences Phila., 1881, p. 451.
Pectunculus lineatus, Reeve.
  Proc. Zool. Soc. London, 1843.
Pectunculus aratus, Conr.
  Am. J. Science, xli, p. 346.
Plicatula ramosa, Lam.
  Anim. s. Vertèbr., vi, p. 176.
Pecten nucleus, Born.
  Mus., p. 107.
Anomia Ruffini, Conr.
  Foss. Med. Tert. U. S., p. 74.
Ostrea Virginica, Gmel.
  Syst. Nat., 3336.

List of species found in the deposits of the Caloosahatchie:

*Murex imperialis,
*   "   brevifrons,
Fusus Caloosaensis,
Fasciolaria scalarina,
*   "   gigantea,
*   "   tulipa,
Melongena subcoronata,
Fulgur rapum,
*   "   contrarius,
    "   excavatus,
*   "   pyrum,
*   "   pyriformis,

Panopæa Menardi,
    "   Floridana,
    "   navicula,
Semele perlamellosa,
*   "   variegatum,
*Rangia cyrenoides,
Venus rugatina,
*   "   cancellata,
    "   Rileyi,
*   "   Mortoni,
*Artemis discus,
*   "   elegans,

*Nassa vibex,
Turbinella regina,
Vasum horridum,
Mazzalina bulbosa,
Voluta Floridana,
Mitra lineolata,
Marginella limatula,
*Oliva literata,
* " reticularis,
Columbella rusticoides,
*Cancellaria reticulata,
Pleurotoma limatula?
Conus Tryoni,
" Mercati?
" catenatus?
Strombus Leidyi,
* " pugilis,
Cypræa (Siphocypræa) problematica,
*Pyrula reticulata,
*Natica canrena,
* " duplicata,
*Crucibulum verrucosum,
Crepidula cymbæformis,
* " fornicata,
Turritella perattenuata,
" apicalis,
" cingulata,
" mediosulcata,
" subannulata,
*Cerithium atratum?
" ornatissimum,
*Bulla striata,
*Siliqua bidentata,

*Dione (Calliste) gigantea,
* " maculata,
Cardium Floridanum,
*Cardium magnum,
* " isocardia,
Hemicardium columba,
*Chama arcinella,
" crassa,
Lucina disciformis,
* " edentula,
* " Pennsylvanica,
* " Floridana,
* " tigerina,
Carditamera arata,
Arca scalarina,
" crassicosta,
* " lienosa,
" aquila,
" plicatura,
" (Arcoptera) aviculæformis,
*Pectunculus lineatus,
" aratus,
Spondylus rotundatus,
*Plicatula ramosa,
Pecten solarioides,
" comparilis,
" Mortoni,
* " nodosus,
* " nucleus,
Anomia Ruffini,
Ostrea meridionalis,
* " Virginica.

The species preceded by an asterisk are living forms.

It will be seen from the above enumeration that of a total of some eighty-nine species forty-one are still members of the recent fauna, constituting forty-six per cent. of the entire number. In addition to these immediate representatives of the recent fauna there are a number of forms which are secondarily representative in the circumstance of being ancestral, *i. e.*, they are manifestly types from which, through an evolutionary process, a portion of the recent fauna has been derived. The close relation which thus binds together the fauna of the present period with that of the Caloosahatchie leaves no doubt as to the Pliocene age of the latter.

The exact position in the Pliocene series which the Caloosahatchie deposits occupy, gauged by the standard of classification adopted by European geologists, cannot be readily determined, owing to the very limited number of forms which appear to be common to both sides of the Atlantic. In the percentage of living forms the formation stands nearest to the Antwerp (Black) Crag, the Diestian of the Belgian geologists and to horizon III of the Bolognese Apennines, in which the proportion of living to extinct forms, as determined by Foresti, is somewhat above 43 per cent. (*vide* Fuchs, *Die Gliederung der Tertiärbildungen am Nordabhange der Apenninen von Ancona bis Bologna*, Sitzungsb. d. k. Akad. Wissensch., lxxi, p. 177, Vienna, 1875; Heilprin, Contributions to the Tertiary Geology and Paleontology of the United States, p. 64, 1884). This horizon constitutes the base of the Italian Pliocene series (Astian) according to those geologists who, like Capellini, recognize in the lower sub-Apennine deposits a transition formation (Mio-Pliocene; Messinian, in part, of Meyer) uniting the Miocene with the Pliocene. The relation held by the "Floridian" formation to the deposits of next oldest date occurring in the eastern United States is almost precisely similar to that which obtains in the case of the Bolognese Apennines. Thus, the "Carolinian" formation, which until the discovery of the South Floridian bed just referred to was considered to represent the newest stage of the Atlantic Tertiary series, occupies a position analogous to the Mio-Pliocene of Capellini. In its upper member, which comprises the beds occurring in South Carolina, the proportion of living to extinct molluscan forms is, as I have elsewhere shown,[*] from 35–38 per cent., and I have indicated that while both stratigraphically and faunally this series is more nearly Miocene than Pliocene, it yet might be considered to occupy a position intermediate between the two. In the upper member of Capellini's Mio-Pliocene, Foresti's horizon II, the percentage of living forms is 38.8. The "Floridian" formation may thus be safely considered to represent the base of the true Pliocene. The percentage of recent forms in the oldest of the British Crag series, is, according to Lyell, upwards of sixty.

[*] Contributions to the Tertiary Geology and Paleontology of the United States, p. 62.

FOSSILS OF THE SILEX-BEARING MARL (MIOCENE) OF BALLAST POINT, HILLSBORO BAY.

GASTEROPODA.

Genus **WAGNERIA**, Heilprin.

I propose this genus for a very remarkable shell, distinguished by peculiarities of structure which broadly separate it from all other Gasteropoda. These peculiarities are: firstly, that the inner or columellar lip is so largely developed as to cause it to envelop a very large, if not the greater, part of the shell, duplicating the outer wall and labrum; and secondly, that through an apparent conjunction of both folds of the mantle, a dome of shell is built over the spire, from which its own walls are separated by a free air-space. This part of the shell appears, therefore, as a second section, completely separated from the basal or apertural division. In what precise manner this dome was formed it is impossible to say, but manifestly the lobes of the mantle must have extended upward from the aperture, arched over, and deposited the shell-layer. The free space which separates the dome from the spire would seem to indicate that the mantle possessed a special rigidity, by which it retained itself. The genus may be briefly characterized as follows:

Shell irregularly oval or rounded-fusiform, intumescently knobbed; spire elevated, broadly scalariform, concealed in a pointed dome which is formed over it by a free upward extension of both lobes of the mantle; aperture narrow, deflected forward in its upper course, where it is reduced to a mere slit, appressed to the body of the shell by a pseudalar expansion of the outer lip; inner lip developed to a most extraordinary extent, covering by its expansion almost the entire, or the whole, shell, duplicating the outer lip.

This extraordinary genus of shells, which I take pleasure in naming after the late Prof. William Wagner, the generous founder of the Wagner Free Institute of Science, of this city, is apparently a near ally of Orthaulax of Gabb (Proc. Acad. Nat. Sciences Phila., 1872, p. 272, pl. ix, figs. 3, 4; Trans. Am. Phil. Soc., xv, p. 234), a form evidently closely related to some of the Rostellariæ, as Calyptrophorus and Hippochrenes (Macroptera), in which the inner lip is frequently abnormally developed. The remarkable duplication seen in Wagneria, produced by the complete backward prolongation of the labium, which actually overlaps a large, if not the greater, part of the labrum, serves, apart from all other characters, to readily distinguish it from the more nearly related forms of the group. Mr. Gabb remarks that in Orthaulax the "adult shell [is] enveloped over the whole spire by an extension of the inner lip," but adds that the "outer lip [is] apparently sharp and simple." An examination of the specimens deposited by Gabb in the Academy of Natural Sciences shows that the latter part of his statement is incorrect, the outer lip being to a considerable extent duplicated. The Florida species which I refer to Wagneria, exhibits this character in a very striking degree, the backward extension of the labium, as seen on a cross-section some distance from the actual base, being fully as ponderous in structure as the labrum proper, which it overlaps as a very distinct outer layer. The duplication exists over at least two-thirds of the shell. The unique dome which conceals the spire is a character not seen in Orthaulax, and is one of the most anomalous structures found among the Gasteropoda.

Wagneria pugnax, nov. sp. Fig. 36.

Shell irregularly oval, obconical, flattened, the flattened appearance being due to three irregular swellings or knobs, one of which immediately adjoins the anteriorly-directed fissure of the aperture; aperture narrow, projected forward (in its upper course) as a closely compressed fissure, which in a crescentical curve ascends to within a comparatively short distance of the apex of the spire; outer lip? (broken in specimen); inner lip largely developed, completely concealing the whorls of the spire, and duplicating for a very considerable extent the outer lip; spire freely enclosed in a pointed superstructure, or dome, built over it by an extension of the mantle; surface covered with longitudinal lines of growth, which extend continuously from the apex to the base.

Length (of imperfect specimens, lacking probably upward of an inch), 2.7 inches; width, 1.75 inch.

What the precise relationship of the genus represented by this species may be I am not prepared to say.

Zittel (Handbuch der Palæontologie, 1, part ii, p. 260) unites Orthaulax with Hippochrenes, but in doing so this eminent paleontologist appears to have been misled by the rather imperfect diagnosis of the fossil given by Gabb. That its position is near to that genus I believe there can be no doubt.

Murex larvæcosta, nov. sp. Fig. 37.

Shell angulated, obscurely scalariform; varices seven to eight in number, obtusely rounded, direct; whorls moderately angulated on the shoulder, crossed by numerous elevated revolving lines, about ten of which on the body-whorl are much more prominent than the remainder, and show a tendency to become lamellar, especially toward the base of the shell; the spaces between the more prominent lines covered with numerous finer (tertiary) lines, and a median secondary line; aperture somewhat more than one-half the length of shell, the (slightly-deflected) canal about one-half the length of aperture.

Length, 1.6 inch; width, .9 inch.

Murex orispangula, nov. sp. Fig. 38.

Shell strongly angulated, markedly rugose; spire elevated, of about five volutions; varices six (on the body-whorl), sharp, deflected obliquely toward the base of the shell; surface of shell very strongly lined, the lines of three series, primary, secondary, and tertiary; those of the first series about ten on the body-whorl, very prominently elevated on the varices, becoming spinose toward the base of the shell and on the apertural varix; aperture slightly exceeding one-half the length of shell, the very narrowly-contracted canal gently deflected.

Length, 1.6 inch; width, .7 inch.

This species may be readily distinguished from *M. larvæcosta*, which it somewhat resembles, by its narrower outline, the smaller number of and greater sharpness of its obliquely directed varices, and its generally rugose surface.

Murex tritonopsis, nov. sp. Fig. 39.

Shell consisting of about six regularly-convex whorls; varices, three on each whorl, profoundly convex and entirely destitute of spines or lamellar processes; two more or less nodulose costæ between each pair of varices; aperture exceeding one-half the length of shell, the canal deflected, very narrow; surface of shell covered with closely placed, elevated revolving lines, which regularly alternate in size.

Length, 1.2 inch; width, .7 inch.

This species very closely resembles *Murex Mississippiensis*, Conr., from the Vicksburg beds, but may be distinguished by the presence of two sharply-defined costæ between each pair of varices, and in the character of the revolving striæ, which are very much finer and more crowded in the Mississippi fossil.

The young of *M. pomum* somewhat resembles the Florida fossil, but may be readily distinguished by the superior angulation of the whorls and the irregularity of the costation.

Murex trophoniformis, nov. sp. Fig. 40.

Shell having the form of Trophon; whorls about six, sub-angulated superiorly, very convex; varices placed at irregular intervals, four on the body-whorl, the intervariceal spaces with one, or two, or even three secondary costæ; aperture about two-thirds the length of shell, contracted

into a short, sharply-deflected, and open canal; surface of shell covered with numerous alternating, elevated lines.
Length, 1.2 inch; width, .8 inch.

Murex spinulosa, nov. sp. Fig. 41.

Shell elevated, elongated, about equally attenuated to both extremities; whorls strongly angulated superiorly, bearing short, outwardly directed, spines on the shoulder angulation; a row of similar (suprabasal) spines in the siphonal region; aperture about one-half the length of shell, the canalicular portion the longest; umbilicus long and open; surface of shell below the shoulder with a limited number of prominent revolving lines, four on the body-whorl.

Length, slightly exceeding one inch; width, half-inch. This species somewhat resembles the recent *M.* (*Urosalpinx*) *fusiformis* of Adams.

Latirus Floridanus, nov. sp. Fig. 42.

Shell fusiform, about equally tapering; whorls convex, sub-angulated superiorly, costated; about ten obtuse costæ on the body-whorl; aperture somewhat exceeding one-half the length of shell, contracted into a gently-deflected, open canal of moderate length; outer lip striated internally; columellar folds feeble, one or two in number, somewhat oblique; surface of shell covered with rugose revolving lines, alternate in size.

Length, 1.7 inch; width, .6 inch.

An apparent variety of this form, possibly a distinct species, has a somewhat more depressed outline, a more pointed apex, and is generally more rugose in its ornamentation. The columellar folds are more nearly transverse, and three to four in number.

This species appears to be on the whole most nearly related to the recent *Latirus infundibulum*, from which it differs in the greatly reduced spire, and a proportional elongation of the siphonal tract.

Fulgur coronatum, Conr.
Bull. Nat. Inst., p. 187.

A fossil from the Miocene deposits of Maryland.

Fulgur spiniger? Conr.
Journ. Acad. Nat. Sciences Phila., new ser., i, p. 117, pl. 11, fig. 32, as *Fusus*.

A solitary specimen, somewhat imperfect, which differs from the Vicksburg fossil only in the slightly more depressed character of the shoulders of the whorls.

Turbinella polygonata, nov. sp. Fig. 43.

Shell elevated, turreted; whorls abruptly flattened on the shoulder—rendering the spire scalariform—the upper ones gently convex, obscurely noded or costated; body-whorl quadrangular, with a broad, flat shoulder;

the costæ obsolete, resolved into a number (about eight) of shoulder-nodes, which break the circumferential outline into a polygon; aperture greatly exceeding the spire in length; columellar folds three, transverse, situated immediately below the body of the shell; revolving lines of surface feebly defined, almost obsolete on the body-whorls, except on the siphonal tract, where they are well-marked, and of equal significance.

Length (of fragment, lacking probably two-thirds of an inch below, and a third of an inch above), 1.8 inch; width, .8 inch.

Vasum subcapitellum, nov. sp. Fig. 44.

Shell elevated, pagodæform; whorls of the spire about seven in number, coronated and strongly costated, the concentric lines (two or three) below the shoulder prominent, those on the rugose shoulder less distinct; the coronary spines prominent, sharp, and directed outwardly; body-whorl with a single row of sharp basal spines, about six in number, below which are two not very prominent lines, and above, some seven sharply-defined concentric ridges, separated by interstitial finer lines; shoulder of whorls elevated; outer lip strongly lined internally; inner lip well expanded, but leaving a broadly-open umbilicus; columellar plaits three, transverse, the upper the largest; surface of shell covered with rugose lines of growth.

Length, 1.4 inch; width, .7 inch.

This shell very closely resembles the recent *Vasum capitellum*, especially the young of that form, and might at first sight be readily mistaken for that species. It differs in its less foliaceous aspect, smaller size, the elevation of the shoulder (nearly flat in *V. capitellum*), and in the presence of only a single row of basal spines (instead of two). I believe there can be no doubt as to its being the ancestor of the living form.

Voluta musicina, nov. sp. Fig. 45.

Shell cylindriform; spire elevated, of about seven volutions; whorls convex, strongly costated, impressed below the suture, so as to divide the costæ into a double series; costæ very prominent, obtuse, about ten on the body-whorl, crossed at right angles by rather distantly-placed, elevated revolving lines; outer lip with a reflected border; inner lip distinct in its lower half, plicated over its entire extent, the plicæ increasing in size from above downward, nearly transverse in direction; aperture considerably over half the length of shell, narrow.

Length, nearly two inches; greatest width, at about the middle of the shell, slightly exceeding one inch.

The shell bears a very general resemblance to the recent *Voluta musica*, of which it may be considered an immediate ancestor, differing from that form principally in its narrower outline, the depressed shoulder of the body-whorl, and the subsutural impression. There is no trace of coronation. The form is intermediate between Voluta proper and Lyria, perhaps nearer to the latter.

Voluta (Lyria) zebra, nov. sp. Fig. 46.

Shell cylindriform, with an elevated, slightly scalariform spire of about six volutions; whorls costated, the costæ (about twenty on the body-whorl) closely-placed, sharply-defined, oblique, forming a pseudo-coronation on top of the whorls; outer lip greatly thickened on the border, slightly ascending; inner lip irregularly plicated over its entire extent, the three or four basal plicæ much the strongest; aperture somewhat more than half the length of shell, narrow, elliptical, contracted basally into a short open canal; surface of shell, barring the costæ, smooth over almost its entire extent, with a few impressed revolving lines on the base of the body-whorl.

Length, an inch and a quarter; greatest width, .6 inch.

This shell most nearly resembles *Voluta pulchella* of Sowerby, a Miocene fossil of Santo Domingo (Q. J. Geol. Soc. London, vi, p. 46, pl. ix, fig. 4), but may be distinguished by its narrower spire, the greater number (best seen on the spire) and more direct obliquity of the costæ, and the costal coronation on top of the whorls. Exceptionally the costæ are equally crowded in *V. pulchella*, but the regular convexity of the whorls, and the absence of the subsutural coronation, seem invariably to distinguish that form. Much the same characters separate it from *Otocheilus (Fulgoraria) Mississippiensis* of Conrad, from the Vicksburg (Oligocene) group, which is also a narrower shell. In its ornamentation the Florida fossil more nearly approaches the recent *V. Delessertiana*.

Mitra (Conomitra) angulata, nov. sp. Fig. 47.

Shell ovately cylindriform, longitudinally plicated; whorls of the spire very convex, slightly angulated above; body-whorl more prominently angulated; revolving lines absent or obsolete, except from the base of the shell; aperture somewhat exceeding one-half the length of shell; columellar folds four, the upper nearly oblique.

Length, .4 inch; width, .17 inch.

Conus planioeps, nov. sp. Fig. 48.

Shell broadly conical, rapidly tapering toward the base; spire reduced to a minimum, represented in most specimens by an exceedingly gentle rise, crowned by a papilla (apex); whorls about seven, all of them fully exposed on the crown, the shoulders concentrically lined; revolving lines nearly obsolete over the greater extent of the body-whorl, prominent on the basal portion; notch?

Length, 1.4 inches; width of crown, .8 inch.

Very closely resembles *Conus Haitensis* of Sowerby, a Santo Domingo fossil, from which it may be distinguished by its more regularly depressed crown, and the character of its ornamentation. The latter species is so variable, however, that not impossibly the Florida form may ultimately prove to be only a variety, although in the extensive series of specimens contained in the Gabb collection, illustrating Sowerby's species, I fail to find anything which fully agrees with it.

? **Pleurotoma ostrearum**, Stearns.

I identify with this species a small Pleurotoma which appears to differ (?) from the living form only in having the costæ more distantly removed from one another, and possibly also a little more prominent. It very closely resembles *P. abundans*, of Conrad, from the Vicksburg deposits of Mississippi.

Cypræa tumulus, nov. sp. Fig. 49.

Shell completely involute, inflated, very convex, the greatest elevation being immediately back of the apex; the dome abruptly truncated posteriorly, sloping more gradually in the direction of the anterior extremity; aperture narrow, subcentral, slightly flexuous, directed obliquely over the apex; outer lip produced somewhat beyond the inner lip posteriorly, with about twenty-five evenly placed dental plications; columellar surface flattened, the teeth less prominent; surface of shell covered with very fine revolving lines, which, however (in the specimens before me), are only visible in immediate proximity to the aperture; base gently convex.

Length, 1.6 inch; width, one inch; greatest elevation, .9 inch.

This species may be readily recognized by the marked elevation of its dome, which is more pronounced than in the case of any other American species of the genus, except *C. sphæroides*, Conr., from the Vicksburg (Oligocene) beds, in which this character is still more emphasized. The latter species may be distinguished by its globose form, contracted aperture, and the absence of revolving striæ.

Oniscia Domingensis, Sowerby (1850).

Q. Journ. Geol. Soc. London, vi, p. 47, pl. 10, fig. 3.
Gabb, "Santo Domingo," Trans. Am. Philos. Soc., xv, p. 223 (as *Morum*).

A single individual, measuring .7 inch in length, in which the granules are largely wanting on the columellar surface, a condition which, according to Sowerby, also characterizes the young of the Dominican form. Mr. Gabb affirms that this species is "very different from *Oniscia harpula*, Conr., from the Vicksburg Eocene [Oligocene], although Mr. Conrad has asserted their identity." I must admit, however, that an examination of the type of Conrad's species, described in the Journal of the Academy of Natural Sciences for 1848 (p. 119), inclines me to the belief that Conrad's determination is the correct one. The two forms are certainly most intimately related, despite Gabb's assertion to the contrary; the Mississippi fossil has a somewhat higher spire, and a more thickened outer lip, but these distinguishing characters may belong exceptionally to the single individual which in the Philadelphia collection represents Conrad's species.

Natica amphora, nov. sp. Fig. 50.

Shell semi-globular, depressed on the basal surface; spire elevated, of about four volutions, all the whorls deeply channeled along the sutural line; body-whorl about three-fourths the size of the entire shell; aperture semi-lunate, contracted above, effuse below; inner (columellar) border of aperture direct, diagonal; deposit of callus considerable, leaving a long, narrow umbilical fissure; base of shell sub-carinated; surface smooth.

Length, about four inches; greatest width, across the centre of aperture, 3.7 inches.

This species, the largest of the American fossil Naticas, cannot be readily confounded with any of the hitherto described members of the genus. Although in a general way recalling the recent *N. duplicata*, it is immediately distinguished from that form by the deeply impressed sutural-line and the exposed umbilicus. Its nearest ally appears to be *N. maxima*, Grateloup, from the deposits of Bordeaux and Dax, France, but it lacks the peculiar expansion of the body-whorl of that species, and further differs in the exposed umbilicus.

Amaura Guppyi, Gabb.

Trans. Am. Philos. Soc., xv ("Topography and Geology of Santo Domingo"), p. 224.

Identified by a single specimen.

Natica streptostoma, nov. sp. Fig. 51.

Shell depressed, oblique, with the spire almost concealed; aperture very large, sigaretiform, the border flattened on the columellar side, and folded over into a pseudo-carina, which passes beneath the labium as the outer bounding-line of the umbilicus; umbilicus narrow, vertical; surface smooth.

Length (height), one inch; greatest width, diagonally across the aperture, 1.2 inch.

May be readily identified by the large, oblique aperture, and the basal carina.

Turritella pagodæformis, nov. sp. Fig. 52.

Shell gently elevated, gradually tapering; whorls numerous, hollowed medially, with an expanded base, which projects considerably beyond the boundary of the whorl upon which it rests, and forms a series of well-marked carinations; a secondary carination above the basal one, followed (in the direction of the apex) by two prominent, faintly beaded, lines and several less prominent ones in the hollow of the whorl, and these again by several alternately placed lines of less value; aperture quadrangular; base flat.

Length of longest fragment, three inches; greatest width, .7 inch.

Turritella Tampæ, nov. sp. Fig. 53.

Shell moderately elevated, the whorls flattened, slightly impressed in the middle, and becoming discontinuous at about a distance of an inch and a half below the apex; revolving lines distinct on the upper whorls, becoming more or less obsolete on the basal ones, except those in the medial impressed furrow, where they remain distinct, appearing somewhat crowded, and alternate in degree of coarseness; aperture sub-quadrangular; base convex.

Length ?

Turbo orenorugatus, nov. sp. Fig. 54.

Shell moderately elevated, the whorls regularly convex, ornamented with coarse concentric beaded or "roped" lines, which are of unequal sizes, the third and fifth lines below the suture finer than those between which they are placed; the beads or crenulations oblique (inclining downward to the left), becoming very coarse and irregular toward the aperture, and scaly or imbricated on the base; the basal lines of nearly equal width, except the one immediately adjoining the labium, which is of about twice the normal width; umbilicus covered; aperture oval; base convex.

Length (height), 1.2 inch; width of base, 1.4 inch.

Most nearly resembles the recent *T. crenulatus.*

Turbo helioiformis, nov. sp. Fig. 55.

Shell dome-shaped, the whorls convex, closely enveloping above— toward the apex—less so below, ornamented with numerous equally-placed revolving lines, upon one or more of which immediately adjoining the suture there is a faint crenulation; aperture obliquely-oval; base convex; umbilicus deep, round.

Length (height), .4 inch; diameter of base, .6 inch.

Most nearly resembles *Turbo* (*Omphalius*) *viridis.*

Delphinula (?) solariella, nov. sp. Fig. 56.

Shell turbinate, moderately umbilicated; whorls subangular, channeled on the basal margin, ornamented with about five concentric beaded lines, the beads largest on the upper lines; base of shell flattened, indistinctly rayed, with an equal number of revolving beaded lines, the beads most prominent on the umbilical line; aperture orbicular, the border nearly continuous; umbilicus deep.

Length (height), .18 inch; width of base, .2 inch.

Closely resembles *Solariorbis bella*, of Conrad, from the Claiborne (Eocene) sands of Alabama, but the whorls in that shell are much more angular, and have two equally prominent circumferential channels instead of the single basal one seen in the Florida fossil. The generic position of the species cannot be definitely determined.

Genus PSEUDOTROCHUS, Heilprin.

Shell turbinate, umbilicated, with the general aspect of the members of the family *Turbinidæ* or their allies, but differing in the siphonate character of the aperture; aperture round, the lip continuous except at the base, which is truncated through the formation of a sharply and obliquely deflected short canal.

I propose this genus for a rather anomalous shell, whose relationship I cannot even guess at. As stated in the generic diagnosis it recalls in habit the turbos, troques, or delphinulas, from which, however, it is immediately separated by the apertural canal. It also in a measure recalls *Trichotropis*, but is of a much firmer and heavier build. Whether or not the shell was nacreous in structure I am unable to say, as the original material has been completely replaced by silica. I know of no form, either recent or fossil, with which it can be said to be closely related.

Pseudotrochus turbinatus, nov. sp. Fig. 57.

Shell doubly turbinate, sloping about equally to base and apex; whorls of the spire crenulated on the angulation immediately above the suture, concentrically striated; body-whorl sharply angulated and sub-carinated in the middle, the crenulations appearing as pseudo-costulations, which are crossed by several transverse lines; base of shell pyramidally convex, concentrically ridged and lined; aperture sub-rotund, canaliculate; inner lip raised, and forming a border to the umbilical sulcus.

Length, .8 inch; greatest width, .3 inch.

Cerithium præcursor, nov. sp. Fig. 58.

Shell small, slender, of the general habit of the recent *C. muscarum*; whorls about ten, longitudinally plicated and concentrically ridged, the ridges or lines about three on each of the whorls of the spire, five on the body-whorl, which in some specimens exhibits one or more irregular excrescences; aperture oval, oblique, produced into a short canal.

Length, .6 inch.

Differs from *C. muscarum* in lacking the basal carination of that species; from *C. ferrugineum*, apart from differences in the character of the ornamentation, in the form of the outer lip, which is not sub-orbicular.

Genus POTAMIDES.
Sub-genus Pyrazisinus, Heilprin.

I propose to designate under this name certain shells which combine the general characters of Potamides and Pyrazus, differing from the former in the non-canaliculate character of the aperture, and from the latter in the possession of a deep sinus in the labrum; the outer lip is effuse, thickened—much as Cerithidea—and carried completely over to the labium, so as to enclose a round siphonal aperture, as seen in the recent *Pyrazus sulcatus*.

Pyrazisinus campanulatus, nov. sp. Fig. 59.

Shell elevated, rapidly tapering; whorls of spire about ten, convex, obliquely costated, concentrically striated, appearing generally rugose; costæ nearly obsolete on the body-whorl, which is disfigured by one or more (?) prominent excrescences or knobs; outer lip effuse, broadly-thickened on the border, with a deep, nearly parallel-sided, sinus; basal border of labrum extending completely over to the columellar surface, enclosing a round siphonal aperture.

Length, nearly two inches; width of base, one inch.

Partula Americana, nov. sp. Fig. 60.

Shell ovately-cylindrical, of about seven volutions; the whorls very convex, longitudinally finely lined, the lines, which are barely visible to the naked eye, somewhat more regular than simple lines of growth, and directed downwards obliquely to the right; body-whorl nearly two-thirds the length of shell; aperture narrowly oval, vertical, somewhat less than half the length of shell; lip reflected.

Length, .65 inch; width, .3 inch.

This shell, as far as I am aware, is the first fossil species of Partula known, and is remarkable as extending the range of the genus to a region removed by one-half the circumference of the globe from its true habitat. In what manner its ultimate distribution was effected can only be a matter of conjecture. The species is closely related to *P. grisea*.

Helicina, sp.?

Several specimens closely resembling in outline *H. substriata* of Gray.

Strophia, sp.?

One specimen, very like *S. incana* of Binney, only a trifle broader; compared with recent specimens from Florida.

LAMELLIBRANCHIATA.

Venus penita, Conrad.
Am. Journ. Science, second ser., ii, p. 399.
? *Venus Floridana,* Conr., id., ii, p. 400.

Shell cuneiform, evenly rounded anteriorly, produced posteriorly; base sinuous; umbones prominent, overlooking a broadly cordiform lunule; the posterior slope sharply angulated, the angulation preceded by a gentle undulating fold; ligamental margin very oblique, and straight from umbo to extremity; cardinal teeth robust; external surface covered with fine concentric lines, the series interrupted at irregular intervals; base crenulated.

Length of largest specimens, 1.3 inch; height, nearly one inch.

The shell is not produced posteriorly to the extent that is represented in Conrad's figure, which is taken from a cast; nor is the anterior portion prolonged much beyond the beaks, so that despite its peculiar cuneiform outline the shell appears high. I have little doubt that Conrad's *V. Floridana* is the young of this species, which is closely related to the recent *V. macrodon* of Deshayes, from the coast of Central America. The latter form is distinguished by its much coarser ribs, and the interstitial semi-line that appears on the posterior angulation.

Venus magnifica, Sowerby.
Thesaurus Conchyliorum, ii, p. 704, pl. 153, fig. 5; Gabb, "Santo Domingo," Trans. Am. Philos. Soc., xv, p. 249.

A single valve, which is undistinguishable from the Dominican fossil (Miocene) and the recent species of the Philippine seas; it differs from *V. puerpera* in having a straight hinge-line. In the collection of the Academy of Natural Sciences of this city there is an undetermined species of Venus from Egmont Key, Florida, which is very closely related to our fossil. It differs in the want of regularity of the concentric raised lines, and in its broadly cordiform lunule.

? **Cytherea staminea,** Conrad.
Foss. Med. Tert. U. S., pl. 21, fig. 1.

Two valves, which differ in but insignificant details from the Miocene fossil of the Atlantic slope.

? **Cytherea Sayana,** Conrad.
Foss. Med. Tert. U. S., p. 13.

A single valve, which has much the aspect of this species, but is a somewhat longer shell and less convex proportionately. It may possibly represent a distinct form.

Cytherea nuciformis, nov. sp. Fig. 61.

Shell erect, sub-trigonal, moderately convex; base evenly rounded, posterior slope rapidly declining; beaks elevated; surface covered with fine concentric lines of growth, disposed in a somewhat interrupted series; teeth?

Length, .8 inch; height, .7 inch.

Several specimens which can be readily identified by their small size and erect outline.

? Chama macrophylla, Chemnitz.
Conch. Cab., vii, p. 149.
Gabb, "Santo Domingo," Trans. Am. Philos. Soc., xv, p. 251.

Numerous small shells, the largest not measuring over one inch in greatest extent, which have a general resemblance to the recent form. In the absence of larger specimens I prefer to consider the identification as somewhat doubtful, seeing how very closely the young of different species of Chama resemble one another. The species appears to be both dextral and sinistral, unless, indeed, two distinct forms are represented by the valves in my possession. One or more of the individuals are undistinguishable from Conrad's *Chama congregata* (Miocene of the Atlantic border).

Lucina Hillsboroensis, nov. sp. Fig. 62.

Shell (known only by the left valve) disciform, suborbicular, evenly rounded anteriorly and basally, truncated posteriorly; beak pointed, sub-central; the pre-umbonal border rapidly declivous, direct; two oblique, fairly prominent, cardinal teeth; surface covered with numerous regularly-placed, concentric and slightly flexuous, lamellæ, about eighteen to the inch, between which are seen finer lines.

Length, 2.2 inches; height, the same.

Differs from *Lucina disciformis*, Heilpr., in its suborbicular outline; from the recent *L. filosa*, apart from other characters, in lacking the convexity of that species.

Crassatella deformis, nov. sp. Fig. 63.

Shell thick in substance, obliquely-oval, the beaks well anterior; anterior border, beginning at the beaks, evenly rounded; posterior border abruptly truncated; basal margin evenly rounded, not flexuous, crenulated; external surface profoundly sulcated, the sulci not extending beyond the angulation of the broad posterior slope, which is slightly hollowed, and only shows the lines of growth.

Length, nearly three inches; height to summit of umbo, 1.7 inch.

This species can be readily recognized by its oblique form, the broad posterior slope, and the prominence of the sulcation.

Cardita (Carditamera) serricosta, nov. sp. Fig. 64.

Shell ventricose, obliquely-oval, highest in the anterior region; umbones well anterior, very prominent, overlooking a deeply impressed, cordiform lunule; hinge-tooth (in right valve) an elongated lamellar plate, which advances beneath the lunular depression (where it is thickened), and is received into a corresponding sulcus in the left valve; external surface radiately ribbed, the ribs about sixteen in number, profoundly elevated, narrow—much narrower than the interspaces—and strongly knobbed or serrated, those of the posterior slope irregular in size; base creno-carinated.

Length, 1.3 inch; height to the top of umbo, one inch.

This species bears a close resemblance to the recent *C. laticostata*, but may be distinguished by the narrowness of its ribs—as broad as or broader than the interspaces in the recent form—and the prominence of its umbones.

Area imbricata, Bruguiere.
Encycl. Méth., 1789, p. 98.
Gabb, "Santo Domingo," Trans. Am. Philos. Soc., xv, p. 254.

A number of individuals, which are practically identical with the recent forms from Key West, Fla. (from the collections of Hemphill), and the Miocene fossil of Santo Domingo (*Arca Noæ*? of Guppy, Q. Journ. Geol. Soc. London, xxii, p. 293). The species is also very closely related to, if not identical with, the Mediterranean *A. tetragona* of Poli. *Arca ocellata*, Reeve, from the coasts of the Malay Peninsula, so nearly resembles the Florida fossil as to be barely distinguishable from it. The only points of difference appear to be a more pronounced angulation (in the eastern shell) of the posterior slope, and the lack of radiating lines on the basal portion of this slope. *Arca protracta*, Conrad, from the Oligocene deposits of Vicksburg, is a close ally, but is a much more elongated shell, and has the posterior border emarginated or sinuous, instead of direct.

Area Listeri ? Philippi.
Abbild. und Beschreib. Conchyl., iii (1851), p. 87.

I have identified with this form a number of arks undistinguishable from a recent species of the South Florida coast, which Mr. Tryon has determined to be Lister's species. I am not absolutely satisfied as to the correctness of this determination, since the recent Florida shell lacks the peculiar light color stripe which, according to Philippi's description, would appear to be characteristic of his species, and has the umbonal region in addition less inflated. The general habits and other characters are, however, the same in both forms. Gabb's *Barbatia Bouaczyi*, from the Miocene of Santo Domingo, appears to be identical with the Florida form.

Area aroula, nov. sp. Fig. 65.

Shell moderately elongated, sharply angulated on the posterior slope, the dorsal and ventral borders nearly straight and parallel with one another; dorsal (hinge) line not much more than half the length of shell; anterior border projecting forward basally; posterior border acutely angulated with the base; beaks anterior, not very prominent, nor very widely separated; ligamental area narrow; teeth almost obsolete in the middle of the hinge-line, becoming oblique toward either extremity; interior of shell deep; external surface closely ribbed, the ribs strongly imbricated by the rugose lines of growth; ribs most prominent on the posterior slope, where they are echinated.

Length, 1.7 inch; height to top of umbo, one inch.

Leda flexuosa, nov. sp. Fig. 66.

Shell subequal, the posterior portion somewhat the longest; basal margin evenly rounded, not sinuous; posterior or ligamental slope feebly arched, nearly direct; teeth crowded, v-shaped; external surface covered with concentric, not very fine, lines, which are gently angulated and flexed on the posterior slope.

Length, .55 inch; height, .25 inch.

This shell most nearly resembles the recent *Leda costellata* of Sowerby, but differs from that species in the non-flexed basal outline, and in lacking the very pronounced angulation of the concentric lines on the posterior slope. From *L. acuta* it differs in the comparative coarseness of its ornamentation, its larger size, and the posterior flexion in its lines.

Lithodomus, sp.?

Two casts, very much like *L. inflatus* or *L. corrugatus.*

? Lima scabra, Born.
    Mus. Cæs., p. 110.

Two valves which are undistinguishable from the less spinose variety of the recent species inhabiting the West Indian seas. The echination is very fine, appearing somewhat like a raised tessellation. Possibly this form may represent a variety of the East Indian *L. tenera*, of Chemnitz.

List of Species occurring in the Miocene deposits of Ballast Point, Hillsboro Bay.

Wagneria pugnax,
Murex larvæcosta,
" crispangula,
" tritonopsis,
" trophoniformis,
" spinulosa,
Latirus Floridanus,

Turbo heliciformis,
Delphinula (?) solariella,
Pseudotrochus turbinatus,
Cerithium precursor,
Potamides (Pyrazisinus) campanulatus,
Partula Americana,

Fulgur coronatum,
" spiniger?
Turbinella polygonata,
Vasum subcapitellum,
Voluta musicina,
" (Lyria) zebra.
Mitra (Conomitra) angulata,
Conus planiceps,
*Pleurotoma ostrearum,
Cypræa tumulus,
Oniscia Domingensis,
Natica amphora,
" streptostoma,
Amaura Guppyi,
Turritella pagodæformis,
" Tampæ,
Turbo crenorugatus,

Helicina sp.?
*Strophia incana?
Venus penita,
* " magnifica,
Cytherea staminea?
" Sayana?
" nuciformis,
*Chama macrophylla?
Lucina Hillsboroensis,
Crassatella deformis,
Carditamera serricosta,
*Arca imbricata,
* " Listeri,
" arcula,
Leda flexuosa,
*Lithodomus inflatus?
*Lima scabra.

The species preceded by an asterisk are living forms.

Of the forty-seven species here enumerated from four to eight are living forms, so that the representation of the recent fauna might perhaps in a general way be assumed to be about 13–15 per cent. The Miocene age of the deposit is thus placed beyond question; and if the proportion of living forms determined for this limited collection be assumed to be approximately correct for a more extended series, then manifestly the exact position of the horizon will be not far from the base of the Miocene. This accords well with the location of the formation, and its own special faunal relationship. None of the fossils—possibly, with one exception—appear to be identical with forms found in the Oligocene deposits of the southern United States; on the other hand, some six or more—*Oniscia Domingensis, Amaura Guppyi, Venus magnifica, ?Chama macrophylla, Arca imbricata, ?Arca Listeri, Lithodomus,* sp.?—are common to the deposits of Santo Domingo. In these deposits the proportion of recent to extinct forms is claimed by Gabb to be as high as 30 to 33 per cent. (" Topography and Geology of Santo Domingo," Trans. Am. Philos. Soc., xv, p. 101), which would make the formation of considerably newer date than is indicated by the Florida fossils. I have not had an opportunity to verify Mr. Gabb's determination, but from a casual examination of his collection it appears to me that strong exceptions might be taken to many of the specific determinations. Comparisons with a number of forms satisfy me that in at least several cases the selected distinctive characters cannot be relied upon, being more imaginary than real, and this criticism applies as well to cases of specific identification as to those of specific separation. But with all necessary allowances for imperfections and deficiencies, it would still be impossible to determine whether the percentage of recent forms ought rather to be increased or diminished, unless a critical re-examination of all the species were entered into. It is, however, a significant fact, that the percentage, as determined by geologists who preceded Mr. Gabb, is placed very much lower than by Gabb himself. Thus, by Guppy the proportion is reduced to 20 per cent,, and by Carrick Moore to from 17 to 8 or 9 per cent. (Q. Journ. Geol. Soc. London, xxii, p. 577). Mr. Guppy further recognizes the proportion of living forms among the Jamaican fossils, nearly all of which are stated by Gabb to occur also in Santo Domingo, to be likewise 20 per cent., but in all these cases the material upon which the determination was made was much less complete than that which served as a basis for Gabb's computation, so that not unlikely the latter's figures are more nearly correct than those furnished by his predecessors. Granting the accuracy of Mr. Gabb's conclusions, the Santo Domingo formation would then seem to represent a horizon somewhat higher in the Miocene scale than is represented by the Florida deposits, in which, as has already been shown, the proportion of recent forms is reduced to 13-15 per cent. This conclusion is in a measure borne out by the comparatively limited number of forms that are held in common by the two series of deposits, a fact significantly emphasized when the proximity to each other of the two areas under discussion is taken into consideration. Still, it is not safe to premise on too scanty material, and while it may be admitted without reservation that the silex-bearing deposit of Ballast Point is of Miocene age, its exact horizon in the Miocene scale may be considered to be as yet undetermined, although the strong probability points to its representing a part of the "Virginian" series. It is surprising that so few of the distinctly Miocene fossils of the Atlantic border should be found here, the more especially as on the Big Manatee River, not more than some thirty miles distant (almost due south), such fossils—*Pecten Madisonius, Pecten Jeffersonius, Venus alveata,* etc.—are prominent by their abundance.

The fact that the silex-bearing deposit of Ballast Point can be shown to be unequivocally of Miocene age is important as bearing directly upon the age of the foraminiferal rock occurring at the same locality, and at Magbey's Spring, about a quarter of a mile above Tampa, on the Hillsboro River. It will be remembered that this rock was correlated by Conrad with the white limestone of the Vicksburg (Oligocene) group, and merely from the circumstance of its containing in abundance the remains of a foraminifer, supposed to be a nummulite (*Nummulites [Assilina] Floridanus*). This supposed nummulite is, however, no nummulite at all, but an orbitolite, so that whatever inference may have been drawn from the occurrence of a form considered to be nearly related to the foraminiferal exponent of the Vicksburg beds counts for naught, although in itself the presence in great quantity of an orbitolite would, if not exactly indicate, at least suggest, the Oligocene period. But the genus is also fairly abundant in the periods preceding and succeeding—*i. e.*, Eocene and Miocene—so that corroborative evidence of one kind or another is needed before we can definitely assign its true position as a constituent of rock masses. Now, it is a significant circumstance that the Oligocene rock proper of the Floridian peninsula—that which I have indicated as the "Orbitoitic"—which is characterized by an abundance of remains of the genera Orbitoides and Nummulites (either of the one or the other, or of both), is wholly wanting in the genus Orbitolites, at least no indications of that genus have as yet come to light there. On the other hand, the genus is represented in the Miocene deposits of the island of Santo Domingo, and by a form which differs but little, if at all, from that which is so abundantly developed in the cream-colored or yellowish limestone of Ballast Point and Magbey's Spring. This form appears to be closely related to, if not identical with, *Orbitolites complanata*, a well-known fossil of the European Tertiaries, whose range extends from the base of the Eocene possibly to the present time. Again, in the orbitolite rock of the localities just referred to, I failed to detect even as much as a trace of either Nummulites or Orbitoides, a circumstance of no little significance when the proximity of this formation to the recognized Orbitoitic of the North is taken into account. The conjunction of these circumstances leads naturally to the supposition that the rock in question is *not* a member of the Oligocene series, as has been very generally supposed. Its geographical position, and the fact that the genus Orbitolites is a member of the Dominican fauna, lends strong support toward considering the true age as Miocene, a conclusion which receives further confirmation from the evidence carried by the fossils associated with Orbitolites. These are in most cases in the form of casts and impressions, mainly undeterminable, but a few of them are sufficiently distinct and characteristic to permit of definite location. One of these, and possibly the form that is most abundantly represented, is *Venus· penita*, from the casts and impressions of which in this rock the species was originally described by Conrad. This shell figures very prominently among the silicified fossils of Ballast Point, but is, as far as I am aware, entirely wanting in the Cerithium rock of the Hillsboro River, which, as has already been shown, underlies the rock containing Orbitolites. Other species apparently identical with forms occurring in the silex-bearing "marl" of this locality are *Cytherea staminea* and *C. nuciformis*. A large cone, possibly identical with *Conus planiceps*, is represented by several casts.

It is to be further remarked, that the Cerithium—*C. Hillsboroensis*—

which constitutes the distinctive faunal feature of the underlying cherty-rock of the Hillsboro, and of the tough blue rock which crops out at Ballast Point, is wholly absent from the rock with orbitolites; similarly, the orbitolite appears to be wanting in the Cerithium rock. What the precise age of the latter deposit may be cannot be determined from its faunal features alone, since the Cerithium, which, as far as my own experience goes, constitutes the only clearly definable species among the numerous molluscan impressions, has thus far not been met with in any other formation, and consequently gives no clue as to the horizon represented by it; but from the position occupied by the rock—stratigraphically underlying the Miocene (probably the lowest member of the Miocene) and geographically wedged in between the Oligocene and Miocene—from both of which it differs widely in faunal characters—I think it may be fairly assumed that it lies on the border horizon of the two series, forming the transition ground.

FOSSILS FROM LOCALITIES NORTH OF BALLAST POINT.

Cerithium Hillsboroensis, nov. sp. Fig. 67.

Shell elevated, of ten or more volutions; sutures impressed; whorls ornamented with four clearly-defined lines of granulations, the granulations of the top series very large, prominent, and somewhat in the form of tubercles; those of the second line very minute; moniliform and nearly equal on the third and fourth lines, in some cases those of the third line most prominent, in other cases the reverse; surface covered with longitudinal, curved creases; base depressed, with some four or five revolving lines; aperture?

Length, 1.5 inch.

Of the type of the European *Cerithium elegans*, but the moniliations on the lower lines of the whorls are direct, and not oblique, and the number of such lines is also different; the upper granulations are, in addition, comparatively more prominent.

Very abundant in the rock forming the bed of the Hillsboro River, which is the first example in this country of a true Cerithium bed. The horizon represented is probably the junction of the Oligocene with the Miocene.

Cerithium cornutum, nov. sp. Fig. 68.

Shell elevated, rapidly tapering, of about 10–12 volutions; whorls convex, strongly costated, the costæ (about seven on the penultimate whorl) oblique and somewhat sigmoidal; body-whorl with two (or three?) broadly-spreading prominences or horns, one of which is situated obliquely over the aperture, partially bounding the posterior siphonal canal; aperture oblique, terminating in a short deflected canal; inner lip broadly-reflected, partially ensheathing the apertural horn.

Length (of imperfect specimen, lacking probably a half-inch, or more), 1.8 inch; diameter of base, .8 inch; length of horn, .3 inch.

From the Oligocene (?) formation of the Pithlachascootie River, a short distance above the mouth of that stream; obtained by Mr. Willcox. This form may be readily recognized by its peculiar cornual protuberances.

Orbitolites Floridanus, Conrad (sp.)

    Am. Journ. Science, new ser., ii, p. 293, as *Nummulites (Assilina)*.
    Nemophora Floridana, Conr., Proc. Acad. Nat. Sciences Phila., 17, p. 74, 1865.
    Cristellaria? Floridana, D'Orbigny, Prodrome de Paléontologie, ii, p. 406.

In my paper "On the Occurrence of Nummulitic Deposits in Florida, and the Association of Nummulites with a Fresh-water Fauna" (Proc.

Acad. Nat. Sciences Phila., July, 1882; Contributions to the Tertiary Geology and Paleontology of the United States, 1884, p. 80) I call attention to the vague description and apparently imperfectly represented figure of the fossil which Conrad refers to Nummulites, remarking that its reference appeared to me very doubtful. Up to that time I had not seen any specimens of the fossil in question, my search among rock fragments that had been sent to me by different parties from Florida proving in all cases ineffectual. At Ballast Point, on Hillsboro Bay, and again in the rock at Magbey's Spring, about a quarter of a mile above the town of Tampa, on Hillsboro River, I was fortunate in finding great quantities of the form that I had been so long in search of, and which had been overlooked for a period of nearly forty years. A cursory examination of the species immediately confirmed my suspicions as to the inaccuracy of its generic determination. The species does not even belong to the great group which includes Nummulites, much less to the genus; it is a true orbitolite, and very close specifically to—if not, indeed, identical with—the common European *Orbitolites complanata*. Its internal structure can be determined even with an ordinary hand-magnifier with considerable precision. The greater number of the individuals are regularly involute, but others assume the cycloidal form represented by Conrad, an appearance in some cases brought about by an irregular exposure of the different planes of the test. More frequently, perhaps, the same form is due to an actual exocyclic involution of the test, as has also been observed by Carpenter and others in the European fossil and the recent species.

The probable Miocene age of the Orbitolitic rock has been commented on in the last section.

Other Foraminifera observed in the peninsula were:—

Nummulites Willcoxi, Heilpr.
Proc. Acad. Nat. Sciences Phila., 1882, p. 191; Contributions to the Tertiary Geology and Paleontology of the United States, 1884, p. 80.

Very abundant in the rock at Loenecker's, on the right bank of the Cheeshowiska River, about four miles above its mouth.

Nummulites Floridensis, Heilpr.
Proc. Acad. Nat. Sciences Phila., 1884, p. 321.

Associated with the preceding in the same locality.

Orbitoides ephippium (sella), Schloth.
Die Petrefact., 1820, p. 89.

Very abundant at the nummulite locality on the Cheeshowiska; less abundant near the mouth of the river (John's Island, etc.), at the springhead, and in the rock of the Homosassa.

? **Orbitoides dispansa**, Sowerby.

Trans. Geol. Soc. London, 2d ser., v, pl. xxiv, fig. 10, 1840, as *Lycophris*.

With the preceding.

Heterostegina, sp.?

In the Miliolite limestone of the Homosassa River.

Sphæroidina, sp.?
" " " " "

? **Bilooulina**, sp.?
" " " " "

Triloculina, sp.?
" " " " "

Quinqueloculina, sp.?
" " " " "

Spiroloculina, sp.?
" " " " "

Several of the last named genera are also represented in the rock near the head-springs of the Cheeshowiska River, and in the mass that crops up on the immediate ocean-front above the landing at Clearwater.

The following table exhibits the relations of the Tertiary formations of the eastern and southern United States :—

Atlantic and Gulf Tertiaries of the United States.

			Foreign Equivalents.
POST-PLIOCENE.			
PLIOCENE.	FLORIDIAN.	Deposits of the Caloosahatchie.	Astian, in part ; Foresti's horizon III of the Bolognese sub-Apennines?
MIOCENE.	CAROLINIAN. (Upper Atlantic Miocene.)	Deposits of North and South Carolina ("Sumter" epoch of Dana). Fossiliferous beds of Rocky Bluff, Manatee River, and of Philippi's Creek and Little Sarasota Inlet, Florida?	Probably the equivalent of a portion of the Messinian of Mayer (Sarmatian, in part, of Austrian geologists), and of the Miocene Pliocene of the Bolognese sub-Apennines.
	VIRGINIAN. (Middle Atlantic Miocene.)	Deposits of Virginia and the newer group in Maryland ("Yorktown" epoch, in part, of Dana). Silex-bearing "marl" of Ballast Point ; Orbitolite rock of Hillsboro Bay and River (Florida)?	Probably of the age of the "Second Mediterranean" of the Austrian geologists, and of the faluns of Touraine ; Caroni beds of Trinidad; and Miocene of Santo Domingo, Jamaica and Cumaná?
	MARYLANDIAN. (Lower Atlantic Miocene.)	Older Miocene deposits of Maryland, and possibly the lower beds in Virginia ("Yorktown" epoch, in part, of Dana).	Probably (or at least partially) the equivalent of the "First Mediterranean" of the Austrian geologists, and of the faluns of Léognan and Saucats.
OLIGOCENE.	ORBITOITIC.	Strata characterized by species of *Orbitoides*. Vicksburg beds, Florida Nummulitic beds, etc.	Aquitanian. Deposits of Crosara and Castel Gomberto (Vicentin), Oligocene of the Mayence basin, sands of Fontainebleau, lower limestone of Malta, Fernando beds on Trinidad, Antigua chert, St. Bartholomew Oligocene.
EOCENE.	JACKSONIAN.	Jackson beds of Mississippi, "White Limestone" of Alabama.	Barton Clay (Bartonian). Sands of Beauchamp?
	CLAIBORNIAN.	Fossiliferous arenaceous deposit of Claiborne, Ala., etc.	Age of the "Calcaire Grossier" of France (Parisian).
	BUHRSTONE.	Beds below the true Claibornian on the Alabama River, "Chalk Hills" of the southern part of the State, etc. "Siliceous Claiborne" (Hilgard) of Mississippi. Maryland Eocene, in part?	Londonian?
	EO-LIGNITIC.	Lignite, sands, and clays situated at the base of the Tertiary series in Alabama, etc. Marlborough and Piscataway beds of Maryland? Shark River deposits of New Jersey.	Thanetian? Bognor rock?

ADDITIONS TO THE FLORIDIAN FAUNA.

Tropidonotus taxispilotus (?) var Brocki. Pl. 17.

I venture to describe, under the above name, the ophidian figured on plate 17, which agrees in general characters with one of the common forms of southern water-snake (*Tropidonotus taxispilotus*), but yet differs in certain elements of structure, which, taken by themselves and under absolutely normal conditions, would be considered to be of at least generic value. This peculiarity of structure rests principally in the disposition of the parietal head-shields, which, instead of consisting of the normal triangular pair, meeting in the median line, diverge from one another, leaving in the opened posterior angle or space a pair of accessory minor plates, that might be termed inter-parietals. The presence of this accessory pair may be due to a want of coalescence in calcification, since even the primary parietals show a disposition to split off into minor plates; or, at any rate, the presence of the outlines of the ordinary rhomb scales in these plates proves them to be composites in structure. In how far the peculiarly modified parietals, and the presence of the accessory pair, may represent permanent structures, I am unable to say, inasmuch as we obtained but a single individual of the species; but it is interesting to note, as will be observed by a reference to plate 17, that the distinctive feature is accompanied by a slight variation also in the arrangement and disposition of the ventral head-shields as well. Recognizing the multiple character of the head-shields, it becomes a question in how far these may be used as a basis for classification. In the present instance, although I have not been able to discover a parallel case, I feel confident that the characters are not of generic, nor probably of even specific value, and I have, therefore, referred the form in question to *Tropidonotus taxispilotus*, although separating it as a sub-species or variety.

Eagle Bay, Lake Okeechobee.

Ictalurus Okeechobeensis, nov. sp. Pl. 18.
(Okeechobee Cat.)

Of the general form and outline of *Ictalurus lacustris*, from which it differs principally in color, the relative position of the dorsal fin, and the greater length of the humeral spine. Head broad, depressed, of nearly equal width and length, with the eye nearly central antero-posteriorly; body moderately stout; dorsal fin nearer to the adipose fin than to the snout (the reverse in *I. lacustris*); humeral process moderately acute, covered by skin, about one-half the length (or more) of the pectoral

spine (barely more than one-third in *I. lacustris*); caudal fin deeply forked, the two lobes nearly equal, with a slight advantage in favor of the upper one. Color above, and largely over the sides, black or bluish-black, yellowish or cream-white on the under surface; one pair of inferior barbels white.
Total length, 21 inches.
Found in Lake Okeechobee.

Aplysia Willcoxi, nov. sp.   Pl. 19.

I would propose this name for a species of Aplysia which is probably fairly abundant in some of the western shallows, although we only met with it in Little Gasparilla Bay. The animal, in its general characters, appears to be most closely related to the European *A. depilans* (*leporina*), with which it may have been heretofore confounded, but differs in several well-marked points of structure, notably in color, the position of the buccal aperture, and in the characters of the pore connecting with the shell cavity. While in *A. depilans*, as described by Rang in his monograph of the Aplysia group (*Histoire Naturelle des Aplysiens*, Paris, 1828), the mouth is placed beneath the tentacular lobes—*i. e.*, the latter are superior, in the Florida species it is central with regard to those organs, the lobes being circumferentially connate, and completely encircling the aperture. The pore leading to the shell-sac is minute, and raised on a small papilla; the stellate markings radiating from the base of the papilla are very feeble, and can barely be discerned without close examination. The shell, which is about two inches in length, is horny-calcareous, deeply emarginate, and striated longitudinally and transversely. General color of the animal sea-green, tinged with purple, and irregularly blotched and speckled with spots of lighter color. Length, 7 to 8 inches. The animal emits a brilliant crimson fluid.

Found on a grass-bank, at a depth of about 2 to 3 feet, and also floating on the free surface of the water.

SUPPLEMENT.

ADDITIONAL SPECIES FROM THE PLIOCENE DEPOSITS OF THE CALOOSAHATCHIE.

For the following new species of fossils I am indebted to Mr. Joseph Willcox and Dr. W. H. Dall, by whom they were collected during a recent visit to the region.

Pecten pernodosus, nov. sp. Fig. 69.

Shell nearly equivalve, strongly plicated and ribbed, the basal margin of both valves incurved; ribs about nine, broadly elevated, and profoundly knobbed on both valves, those of the right valve almost throughout broader than the interspaces, those of the left valve of equal width, or narrower than the interspaces, and alternating in size; knobs closely placed, more or less hollow, about ten on each rib in the largest specimen; ribs and interspaces radiately ribbed or lined, the lines crossed by numerous rugose creases of growth; ears unequal, longitudinally lined or grooved, the lines declivous; cardinal pit moderately deep.

Length, four inches; height, from apex to basal margin, four inches.

This beautiful scallop, which is, with little doubt, the immediate ancestor of the recent *Pecten nodosus*, can be readily distinguished from that species (and likewise from *Pecten subnodosus*, which is hardly more than a variety of *P. nodosus*) by the much greater prominence and regularity of its closely packed knobs, and in the circumstance that both valves are nearly equally knobbed. In *Pecten nodosus* the ribs of one valve, usually the right, are largely destitute of true knobs, although exhibiting here and there ephippial undulations; the knobs are also less regularly rounded, and the radiating lines are less numerous. Much the same differences separate the species from *P. Peedeensis*, from the Miocene of South Carolina.

Cardium Dalli, nov. sp. Fig. 70.

Shell ovately elevated, moderately ventricose, with the beaks apical, touching (or nearly so), and directed slightly backwards; ribs about 30 to 33, smooth, moderately elevated, teretely rounded, with narrow, impressed interspaces; the ribs on the posterior slope narrower and more crowded than over the general surface, minutely echinated in part.

Hinge-line narrow, acutely curved, with prominent lateral teeth; a prominent triangular cardinal tooth in each valve.

Height, from apex to basal margin, 5.3 inches; length (width), 3.7 inches.

This very interesting cockle, which I have the pleasure of naming after Mr. W. H. Dall, the distinguished malacologist of the U. S. National Museum, is closely related to the recent *Cardium subelongatum* from the West Indian seas, of which it is not unlikely the progenitor. In the latter the ribs are much narrower, scarcely exceeding in width the interspaces, and proportionately much more elevated. The echination on the posterior slope in the recent form appears also to be more strongly developed. As far as the color traces remain in the fossil species it would seem that the general scheme of coloring was the same in both species.

An interesting relationship is also maintained between *Cardium Dalli* and the Eastern *C. elongatum*, from the Philippines, which in size and general habit perhaps even more nearly corresponds to the Florida fossil than does *C. subelongatum*; it is, however, a much more ventricose shell.

Cerithidea scalata, nov. sp. Fig. 71.

Shell broadly turreted, scalariform; whorls ten or more, strongly ribbed, those beyond the sixth or seventh whorl from the apex with a more or less hollowed or excavated shoulder; ribs oblique, defined only on the lower half of the later whorls, twenty or more on the body-whorl, with one or two variceal interruptions; revolving lines distinct on the apical portion of the spire, cancellating that part of the shell; aperture? (broken); canal short, moderately deflexed.

Length (of imperfect specimen), 2.4 inches.

Vasum horridum, nov. sp. Fig. 72.

Having received a number of perfect specimens of this beautiful species, I am now able to supplement and complete the description given on page 75 of this report (Fig. 6).

Shell turbinate, thick, with the greatest width at about one-third the distance from the apex to the base; spire moderately elevated, of about 6 to 7 whorls, most of which are doubly coronated or calcitrapated by prominent lamellar or flattened spurs; the spurs regularly increasing in size, with the apices turned slightly backward.

Body-whorl strongly angulated on the shoulder, beautifully coronated, and crossed by about eight prominent revolving ridges, the four immediately following the shoulder coronation nearly equal, scaly, the sixth and seventh, more particularly, carrying long lamellar spines or tubercles, those of the sixth row inflexed upward.

Columellar plaits three, the uppermost by far the most prominent; aperture about two-thirds the length of shell, flexuous inferiorly; umbilicus long and broad.

Cypraea (Siphocypraea) problematica, nov. sp. Fig. 73.

Specimens of this species (*v. ant.*, p. 87, Fig. 12), with a complete coating of enamel, show that the general color of the shell was buff or cream-yellow above, irregularly and minutely spotted with darker shades of the same color (inclining to orange), and impure white below.

Mitra lineolata, nov. sp. Fig. 74.

Specimens of this shell, in certain respects more perfect than the type described on p. 79, indicate that the surface was covered by revolving lines of purple, corresponding in position to the raised lines, and that these were regularly blotched with spots of the same color, resembling the similar markings of *Voluta Junonia*. I have already indicated the characters which doubtfully serve to distinguish this species from Conrad's *Mitra Carolinensis*, and am now more than before inclined to believe that it may prove only a variety of that form.

Conus Tryoni, nov. sp. Fig. 75.

Length, six inches.

The following additional species have been identified as occurring in the " Floridian " (Pliocene) deposits of the Caloosahatchie :

Fusus exilis.	Niso, nov. sp.
Fasciolaria acuta.	Turbonilla, sp.
*Marginella roscida.	Corbula, sp ?
*Terebra dislocata.	*Tellina tenera.
*Columbella lunata.	*Amphidesma equalis.
Cancellaria depressa?	*Semele rosea.
*Conus papilionaceus	*Cardium serratum.
(with color markings).	*Lucina cribraria.
*Conus Floridanus ?	* " radians (Antillarum).
*Trivia pediculus.	*Leda acuta.
*Xenophora conchyliorum ?	Astarte undulata ?
Turbo, nov. sp.	Glandina.
Crucibulum ramosum.	Planorbis.
* " scutellatum.	Amplexa.
*Trochita centralis ?	Paludina.
*Obeliscus arenosus.	

The species preceded by an asterisk are living forms.

With the above were found associated the remains of a proboscidean, horse, alligator and turtle.

Note on the geology of Little Sarasota Bay.—Mr. Willcox furnishes me with the following observations, made during a more recent visit, bearing upon the geology of this region : " Two small fresh-water streams empty into Little Sarasota Bay, not far south of Mr. Webb's house. At the mouth of these streams a bed of ferruginous sandstone has been formed, the largest being about 100 yards in extent along the bay. The iron oxide, cementing the sandstone, undoubtedly was supplied by the fresh-water streams. In this bed are found abundantly many species of shells, such as are now found living in the Gulf of Mexico, one mile distant from this locality. These shells are in good condition, indicating only a small amount of erosion. More than twenty fragments of Indian pottery were found in this bed during a late visit to it, some pieces being nearly as large as a man's hand, and rudely ornamented. Vertebræ and teeth of sharks also abound in this sandstone, also many fragments of manatee bones; the latter were, however, all eroded into smooth, oval forms before they were enveloped in the sandstone."

Mr. Willcox also furnishes the following approximate section of the North Creek exposure, to which reference is made in the report:

Sand (3 feet).

Hard limestone rock (2 feet).

Sand and calcareous marl (6 to 7 feet) containing shells, which are most abundant near the water-level.

Water-level.

The shell deposit, in all probability, belongs to the Pliocene period.

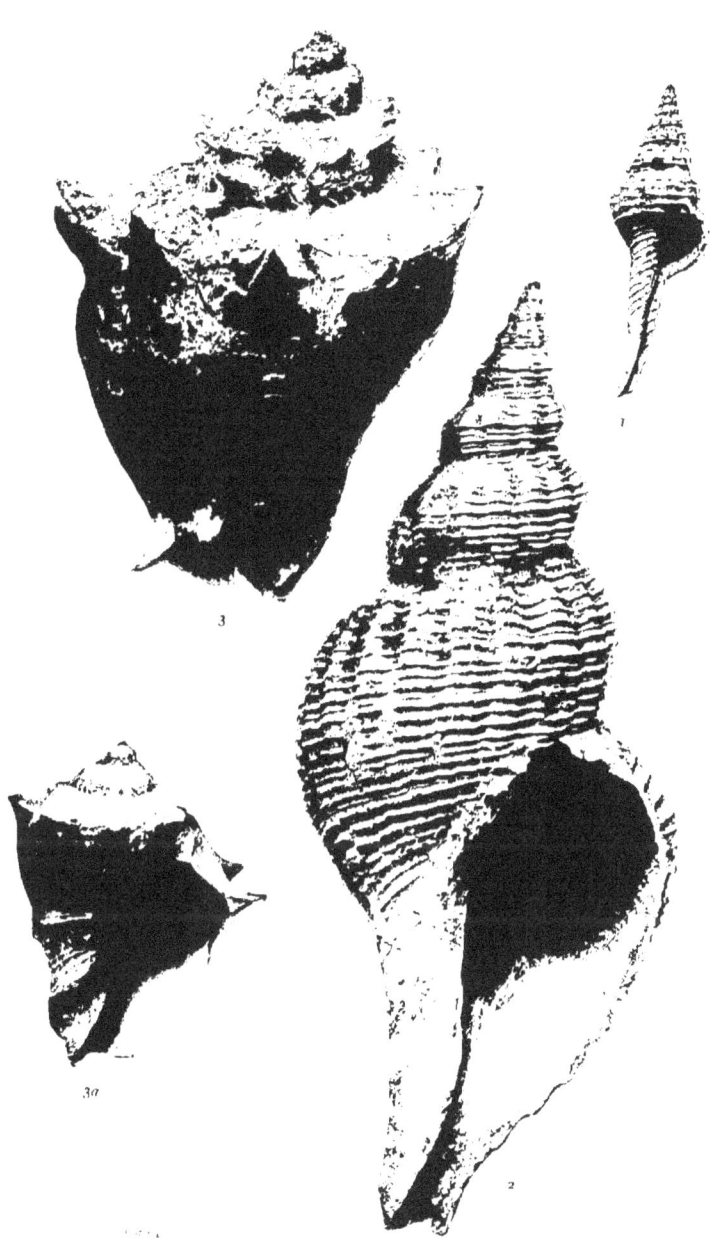

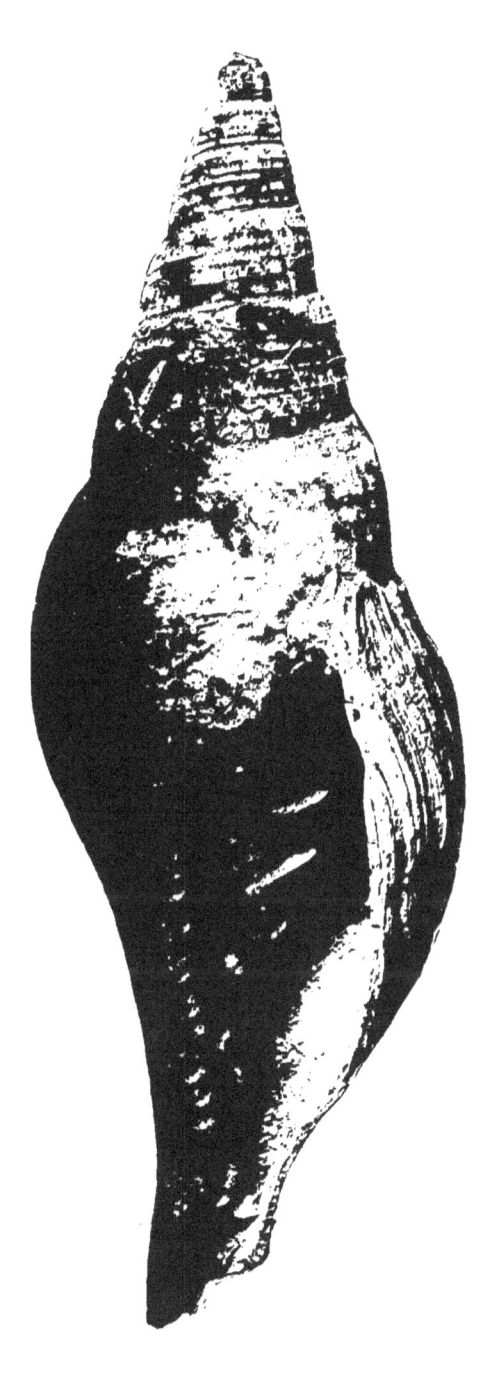

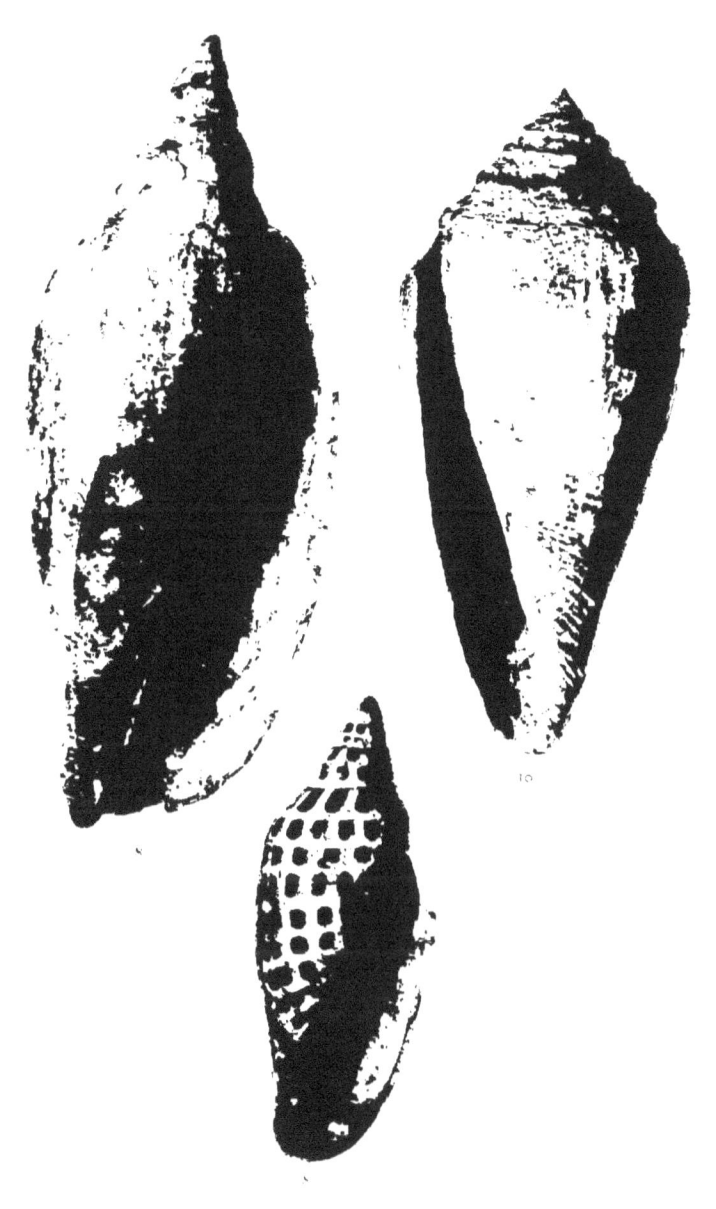

11

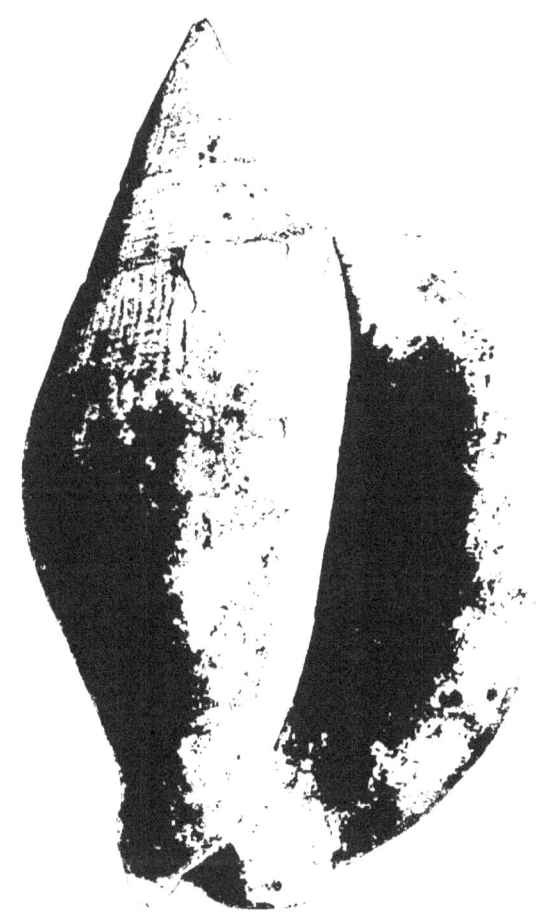

11a

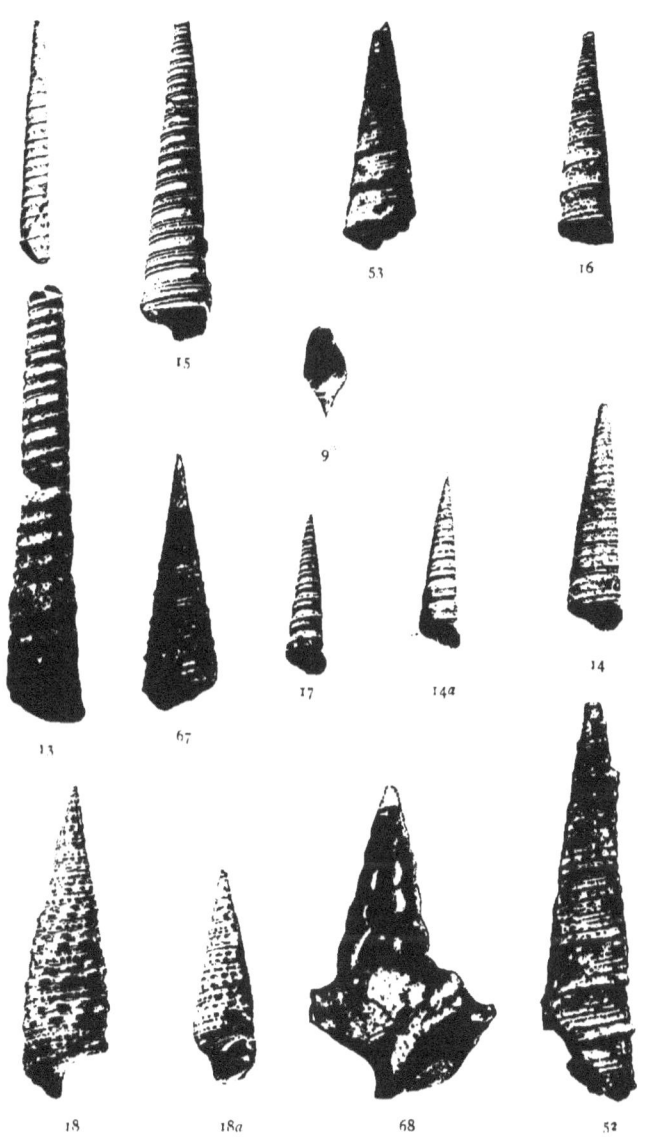

20

19

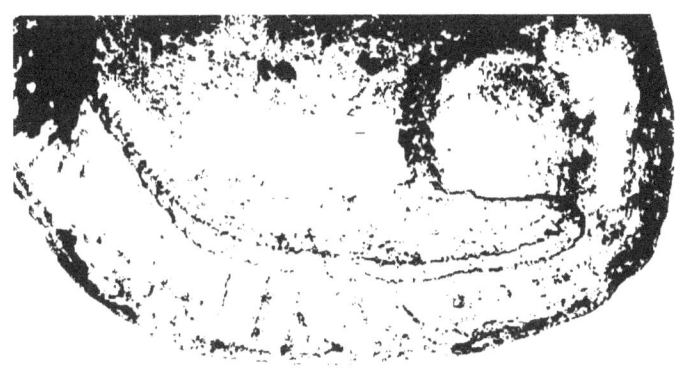

22

21

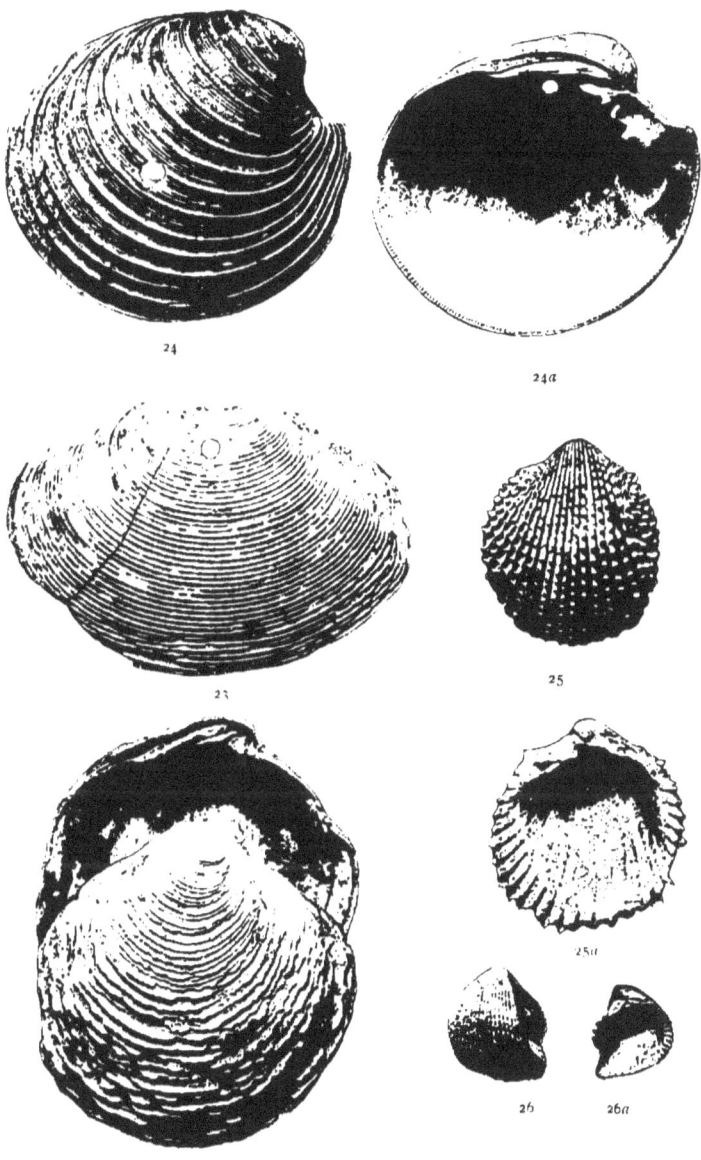

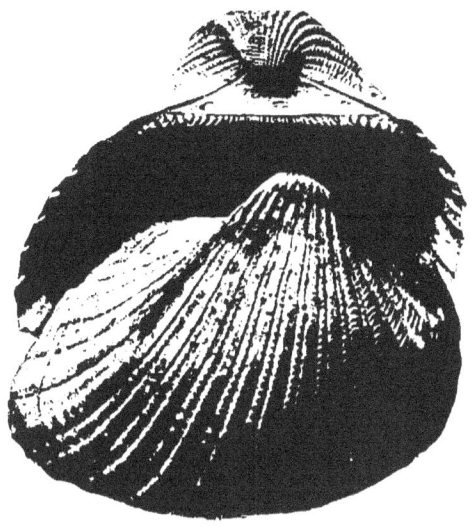

29

31

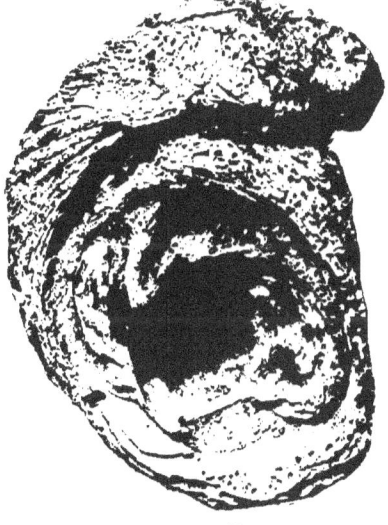

27

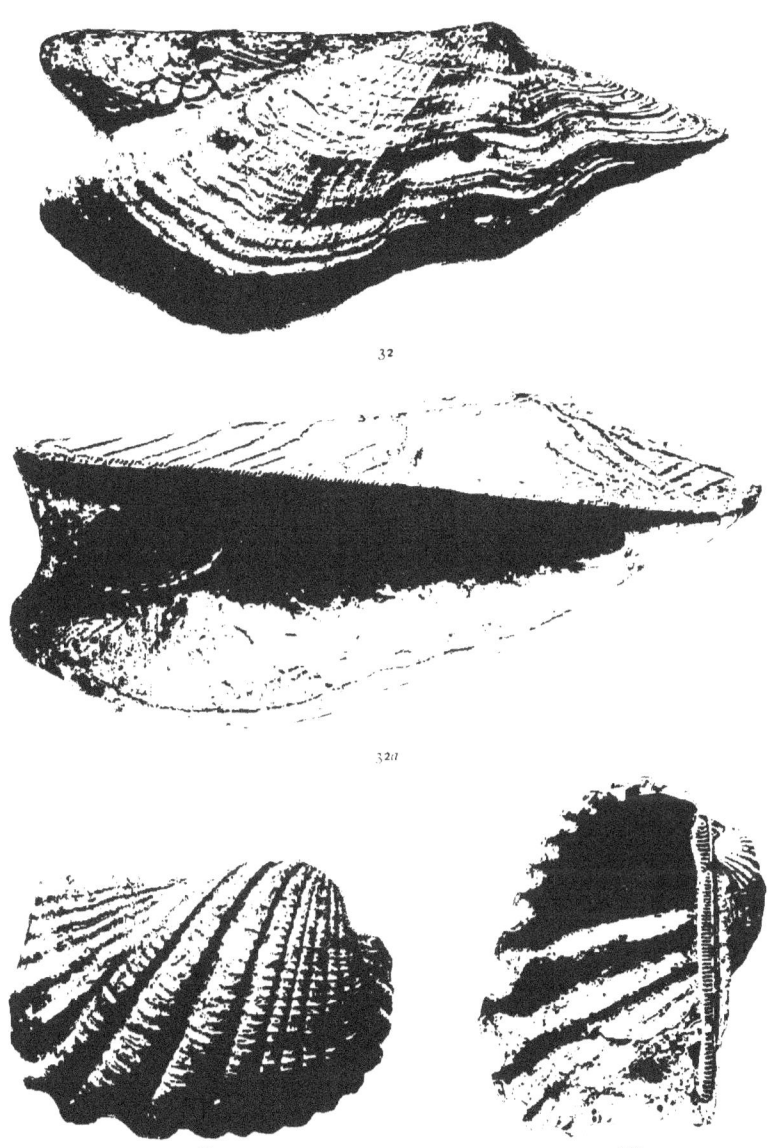

32

32a

30 30a

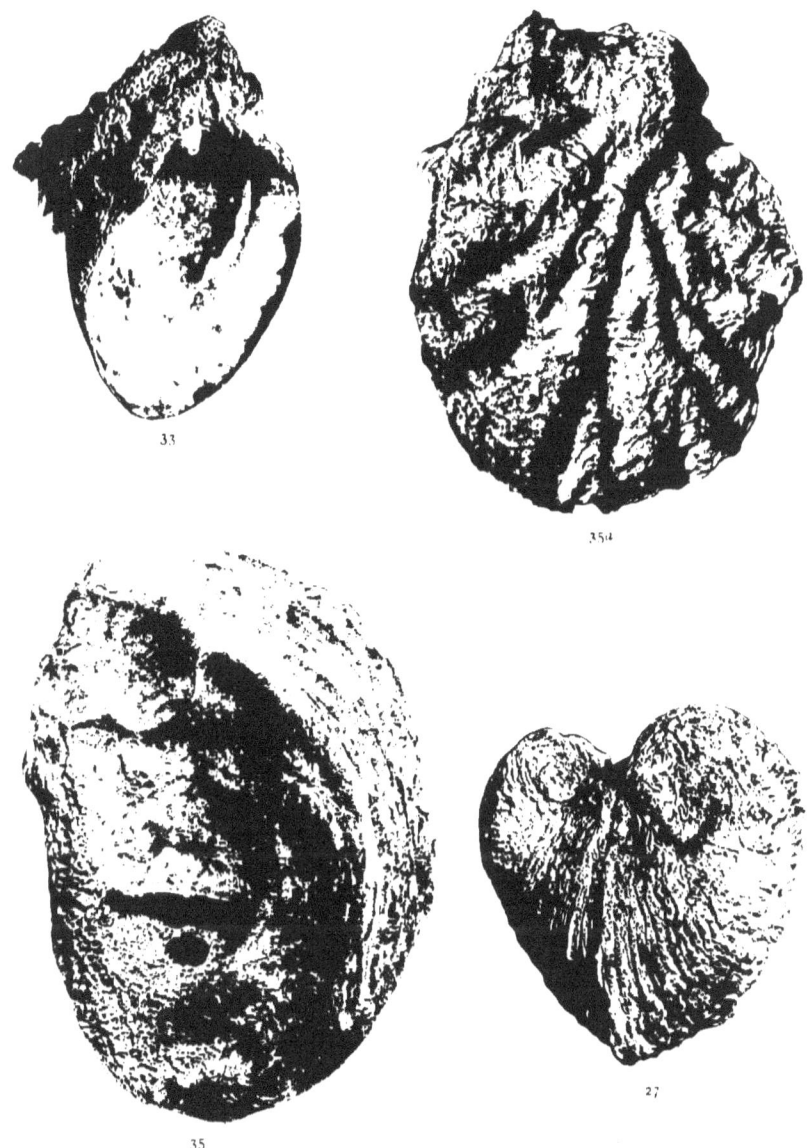

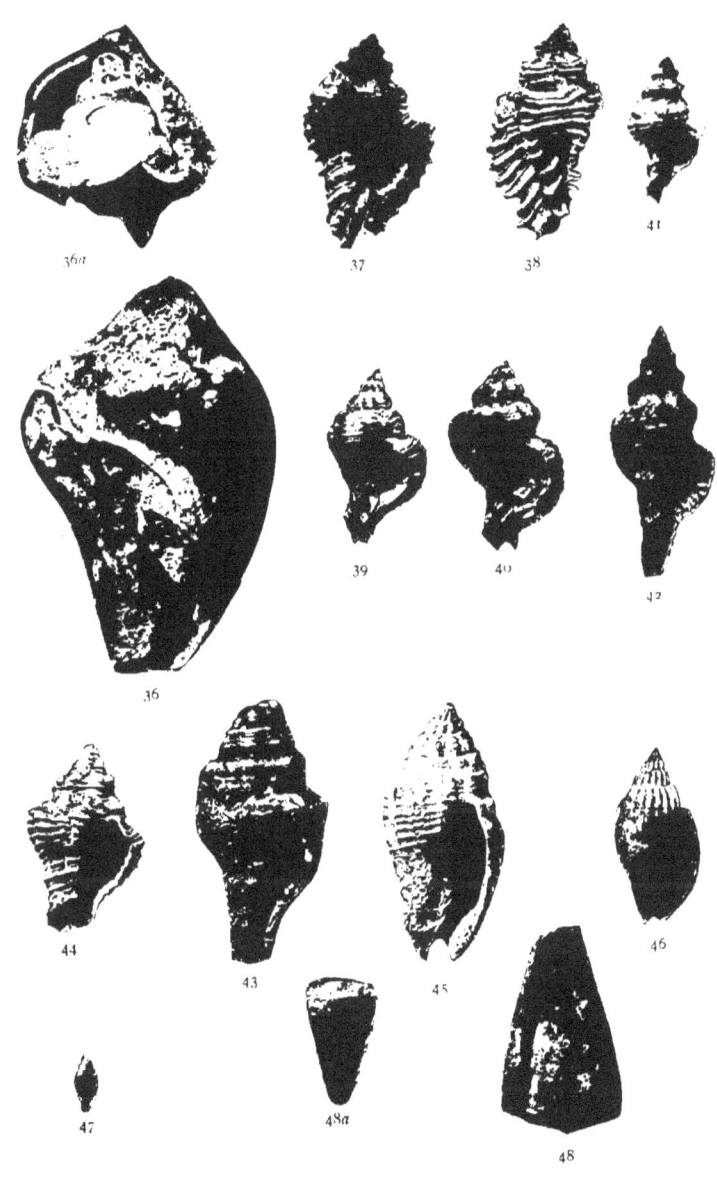

PL. 16

MIOCENE FOSSILS OF FLORIDA

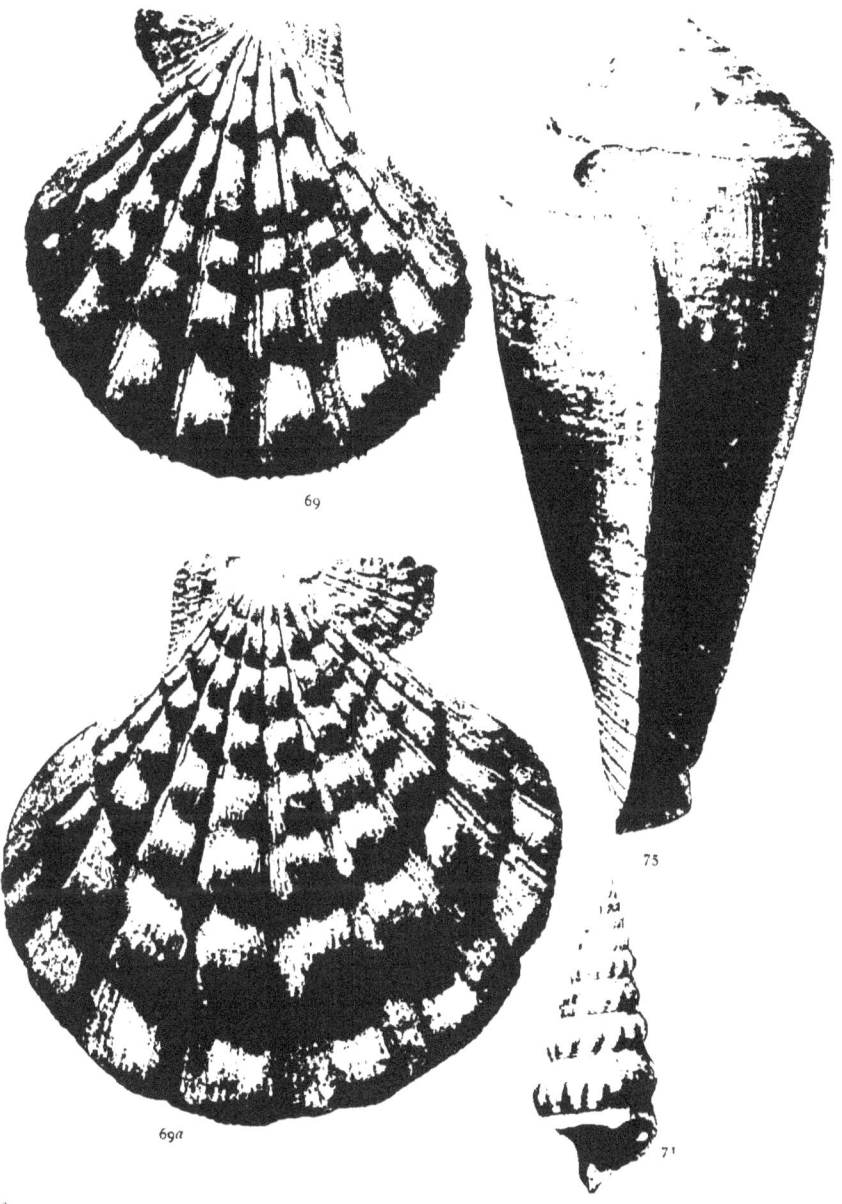

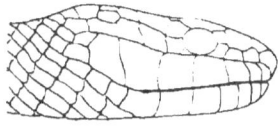
a
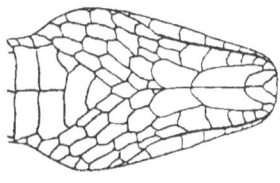
b
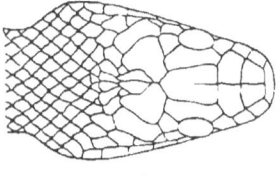
c

Lateral aspect of head. b. Ventral. c. Dorsal

TROPIDONOTUS TAMMILOI J () var. ROOKI

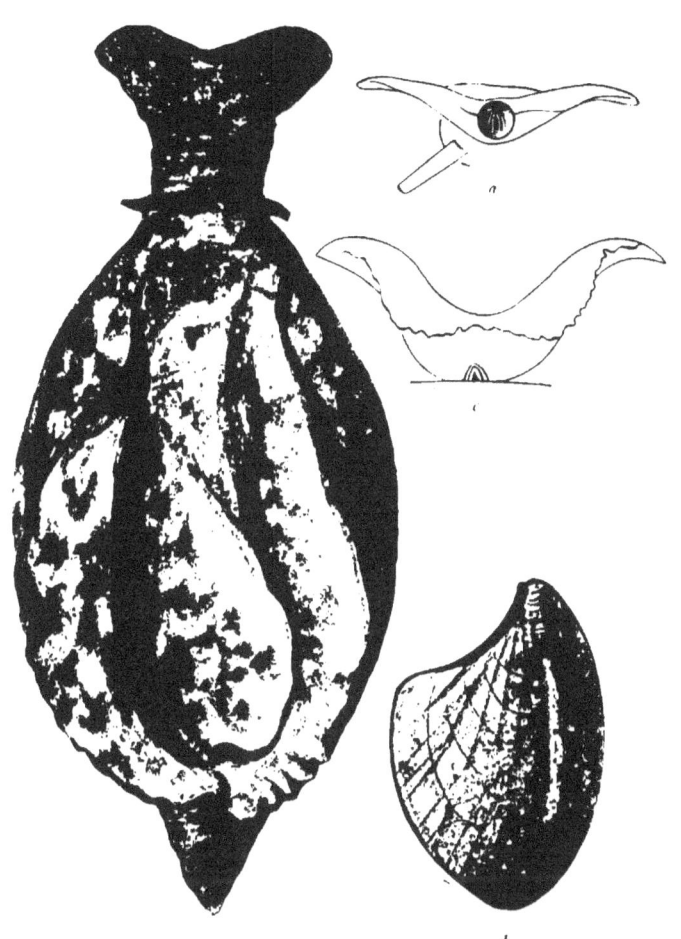

www.ingramcontent.com/pod-product-compliance
Lightning Source LLC
Chambersburg PA
CBHW020826190426
43197CB00037B/714